LighterSide
OF
ADAPTIVE
OPTICS

LighterSide
OF
ADAPTIVE
OPTICS

Robert K. Tyson

PRESS

Bellingham, Washington USA

Library of Congress Cataloging-in-Publication Data

Tyson, Robert K., 1948–
  Lighter side of adaptive optics / Robert K. Tyson.
    p. cm.
  Includes bibliographical references and index.
  ISBN 978-0-8194-7561-9 (alk. paper)
  1. Optics, Adaptive. 2. Imaging systems in astronomy. 3. Optics,
Adaptive--Humor. I. Title.
  TA1522.T96 2009
  621.36'9--dc22

                        2008050742

Published by

SPIE
P.O. Box 10
Bellingham, Washington 98227-0010 USA
Phone: +1 360.676.3290
Fax: +1 360.647.1445
Email: Books@spie.org
Web: http://spie.org

Contents

Foreword

A few months ago I received an email from Tim Lamkins whose official title at the time was Content Acquisition and Development Specialist, Publications for SPIE. SPIE used to be called SPIE—The International Society for Optical Engineering, but now it is called SPIE.

Tim was inquiring whether I would like to write another book on adaptive optics. My immediate reply was "No!"

Tim was persistent, especially with the job title "Content Acquisition and Development *Specialist.*" I told him that there were plenty of good books on adaptive optics already, some of which I wrote myself, and some of which were written by others who copied all my best ideas, but wrote them under their own names. I told Tim that there wasn't much need for another book yet. "Maybe in a few years," I said.

Tim was not taking "No!" for an answer. He explained that the book was to be for a wider audience than the one-billionth of one percent of the population of the world who would normally buy a book on optics. His vision for this book was "for the masses." No math was to be involved.

I was still reluctant. I know Tim was just getting ready to offer a large advance for the book, much like Stephen King would get, I presume, but he didn't need to cut the check ... he told me that it could be "humorous."

His argument was persuasive. I could write a funny book about adaptive optics! So, here it is. If you don't find it greatly informative, technically perfect, and somewhat amusing, tell Tim.

Bob Tyson
November 2008
Charlotte, North Carolina, USA

Chapter 1

Love Is in the Air

It was all up in the air with Barbara and Kenneth.[1] They had been dating for about 20 minutes and Kenneth wanted to take things to the next level, so to speak. Frankly, Barbara did too, but she was thinking that maybe he would just ask her for her cell phone number. They were a perfect couple, actually. They knew absolutely nothing about adaptive optics, nor did they care to know anything. They just wanted to make it to cloud nine[2] and have a blissful arrangement, at least for another 20 minutes or so.

Kenneth looked into his cup of coffee and stirred it gently. He watched the bubbles come to the surface and break. Kenneth wasn't a great thinker. He didn't ask things like: Why do they break when they come to the surface? Or: Is a bubble still a bubble after it breaks?

No, Kenneth was staring into Barbara's azure eyes and thinking: How do I ask her for her cell phone number?

On the other side of the table, Barbara was wondering why her breathing was getting faster. Did she really have an interest in Kenneth? Was it just a crush? Would Kenneth even ask for her cell phone number? She was breathing faster, using precious air. Kenneth was breathless as he stared at her.

---

[1] First of many footnotes. Any resemblance to any persons living or dead is merely a coincidence. Any resemblance to plastic depictions of young adults is only in your imagination.
[2] Or, at least seven.

Without air, we would all be breathless. If we didn't have any air here on planet Earth, we would have to breathe water. The last time I tried that, it didn't work so well. Besides breathing, air is useful for lots of things. It is used to fill balloons. It is used to fill tires, and according to some authorities, will improve my gas mileage if I don't neglect to check it once in a while. Air is why the sky is blue. When air starts moving, it can provide a gentle breeze that cools us from the hot sun on a tropical beach.

But air isn't always good. Sometimes, when it is polluted, it can make our eyes burn. Air can carry nasty little creatures that cause all manner of pestilence. If it gets moving too fast, like around 250 miles per hour in a tornado, it can literally rip our world apart. Turbulent air can throw around even huge jetliners, sometimes to the point of wishing that airports didn't serve pizza.

Though we may enjoy some of the effects of turbulent air, such as twinkling stars, for one class of nocturnal scientists (astronomers), moving air can destroy perfectly good pictures.

Isaac Newton, arguably, the greatest scientist who ever lived and wore, arguably, the silliest wig at Cambridge University, noted in the fourth edition of his book *Opticks,*[3] that "the Air … is in a perpetual Tremor." It is sometimes hard for me to imagine something in perpetual tremor, until I go into a near perpetual tremor when I imagine certain individuals being elected to high levels of public office. Newton, thinking apolitically, was noting how the "only Remedy is a most serene and quiet Air, such as perhaps be found on the tops of the highest Mountains …"

So, since the invention of the telescope in the early 17th century, scientists have been looking for high mountains where they can place their telescopes to avoid the perpetual tremors of the atmosphere. It is an unfortunate irony that a place that doesn't

---

[3] They speld it funy backe then. The full citation is: Sir Isaac Newton, *Opticks: or, a Treatise of the Reflections, Refractions, Inflections and Colours of Light*, 4th ed., William Innys, London (1730).

An artist's depiction of air.

have much air doesn't *have much air*. So what may be a good site for a telescope is also a pretty poor place to take up residence to build and use the telescope.

Fortunately, humans are amazingly adaptable creatures. Evidence of this is all around us. We adapt to the cold air by wearing the skins of creatures that can't run faster than we can or can't hide. We adapt to warm air by sipping cool beverages while covering up with SPF 10,000. Under longer time scales, say, 30,000 years, we adapted to global warming by watching the glaciers recede and form Lake Michigan. Then we built Chicago and hoped for the return of the glacier.[4]

We humans, and I consider myself a reasonable specimen, can adapt to high altitude and sparse air by staying at high altitude and struggling with breathing until our lungs adapt. Distance runners do the same thing, but they wear funny shorts and struggle with breathing for the sheer joy of it.

---

[4] Sorry Bears fans, I could have used Green Bay instead, but chose not to.

Now, with adaptation to high altitude, astronomers are able to build big telescopes on the tops of (some of) the highest mountains. Unfortunately, the problem of turbulence and crummy images does not go away completely. There must be a better way.

An artist's depiction of light rays going through air without any turbulence.

An artist's depiction of light rays going through air *with* turbulence.

In the beginning …

Back in the last millennium, in 1953, Wilson Observatory astronomer Horace Babcock came up with an idea. Although he is often credited with the invention of *adaptive optics*, his design was never put into practice. Remember that 1953 was before spaceflight, the invention of the laser, the personal computer, or Elvis Presley's first hit single. Paris Hilton was a hotel. Babcock suggested how we could measure the turbulence of the air above the telescope and then how we could use a device, similar to a television tube, to compensate for the turbulence. Instead of just relying on higher and higher telescopes in more uninhabitable places, Babcock suggested that we build a device that could ignore the turbulence.[5]

Horace Babcock (1912–2003)
(Photo courtesy of the American Institute of Physics
Emilio Segre Visual Archives, John Irwin Slide Collection.)

---

[5] H. W. Babcock, "The possibility of compensating astronomical seeing," *Publications of the Astronomical Society of the Pacific*, **65**, No. 386, 229 (1953). This is an actual scientific notation, unlike some of the satirical ones that are coming up.

An artist's depiction of Babcock's idea for adaptive optics.

So, did Horace and his buddies go down to the local hardware store in 1953 and buy some lenses and things and glue them together to ignore the atmospheric turbulence? Well, not exactly.[6] To make a long story longer, it took nearly 20 years for anything to get done.

Astronomers and astrophysicists just want to tinker around and take pictures. Their only goal is such a silly one. They want to figure out where we are, how we got here, and where we are going. It's clearly not as exciting as trying to guess what people are going to wear to the Academy Awards.

Scientists needed to develop technology to put Babcock's idea and many other improvements into practice. The world never has enough money to give everybody millions or to make sure nobody gets sick or to provide Botox® treatments for everyone who thinks it's *their right*. To make adaptive optics work, there had to be a reason for it, a big reason.

Along came the cold war. From an American viewpoint, the Soviet Union had thousands of nuclear warheads aimed at our cities, military bases, and nursery schools. From the Soviet viewpoint, it

---

[6] The store didn't have what was needed and a layaway plan really wasn't practical.

probably appeared differently. However, when the advances in laser science progressed to the production of some really high-power beams (hundreds of kilowatts), it was proposed that these intense beams could be used to shoot down an attacking aircraft or missile. There were daunting problems, though. First, a beam powerful enough to melt through hard military targets would also melt the very mirrors necessary to make the high-power beam. Also, the atmosphere was something that could absorb and scatter the high-power beams, removing much of their lethality.

So, with a limited, but large budget, the U.S. Department of Defense began to fund private contractors to solve these problems, develop the technology, and eventually build the complete weapon systems. By the late 1970s, laboratory prototypes were demonstrating the basic functions of adaptive optics components. This was accompanied by proposals for integrated experiments that would be necessary to propel the high-power beams through the unforgiving atmosphere.

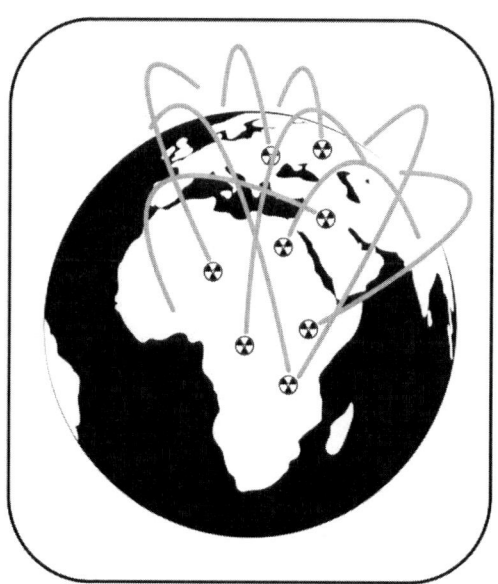

An artist's depiction of a really bad day.

As I look back at the adaptive optics technology of 30 years ago, it seems that it looked quite large and clunky, in contrast to today's technology that is compact and sleek.[7]

By this time, you may be asking yourself: What is adaptive optics? Or you may be asking: What are adaptive optics? Or you may just be wondering, What am I having for dinner?

Is *adaptive optics* singular or plural? Well, frankly my "deer," it is both. *Optic* is usually not used as a singular noun. Webster's Dictionary even points out that the singular noun form, *optic*, is "a pretentiously humorous usage." *Optic* is a very nice adjective to have around for things like optic nerve or optic axis. *Optics* is defined as "the branch of physics dealing with the nature and properties of light and vision." Describing a system of optical things, such as lenses or mirrors, is often called *optics*.

Optics, being a branch of physics, is usually described with a lot of mathematics. Some of the mathematics can be quite daunting. The illustration below shows some of this dauntedness.

$$\Phi(x',y') = \int_{-\infty}^{\infty}\int_{-\infty}^{\infty}\Phi(x,y)e^{-2\pi i(xx'+yy')/\lambda z}dx\,dy$$

$$E = m_0c^2$$

$$4scor+7yrsago-> r\ fa\ \theta rs\ br0t\ 4^{th}$$

Mathematical things that may or may not have anything to do with adaptive optics. You may recognize some of them. Then again, you may not.

---

[7] This is sort of the opposite of my looking into a mirror for the past 30 years.

What is adaptive optics?

It is a premise of this book that the level of math is minimized. So, forget you asked. Let's just concentrate on the bits and pieces of stuff that make adaptive optics *adaptive*. Adaptive optics, being composed of more than one optical element, should be plural. But adaptive optics components become a system, which is singular. If you are talking about the *system*, you ask: What *is* adaptive optics? But, when you are referring to the multiple optical gizmos in the system, you ask: What *are* adaptive optics?[8]

The answer to either of these questions can be "Adaptive optics is a combination of advanced electro-optic technologies used to perform real-time corrections of aberrations to adjust the phase of the light so that images can be improved or a laser beam can be controlled. For example, if we assume a Kolmogorov atmospheric turbulence spectrum, we can show how a Shack-Hartmann wavefront sensor, a digital matrix multiply reconstructor, and a continuous-faceplate deformable mirror can zonally modify the pupil and correct the image."

Or we could simply answer: "The atmosphere screws it up. Adaptive optics unscrews it."

This is a good place to point out that adaptive optics is not used *just* in astronomy to correct for atmospheric turbulence. A bunch of clever people have found ways to use adaptive optics in laser communications,[9] inside laser resonators,[10] for high-energy laser propagation,[11] and for looking inside the human eye.[12]

---

[8] I know that this didn't clear things up at all.

[9] Sending information over a beam of light rather than on a radio wave.

[10] To fix the problems of nonuniform materials, misalignments of optics, and other quantum things that I don't understand and you shouldn't question.

[11] Mentioned earlier as a reason to do anything at all.

[12] Described briefly in Chapter 11, which apparently is a very brief chapter.

The formal scientific definition of adaptive optics. The atmosphere screws it up. Adaptive optics unscrews it.

Meanwhile, Barbara and Kenneth are doing just fine. Their relationship has progressed to the point where Kenneth is actually thinking that he would someday like to marry Barbara. They both see each other as near-perfect, except for a few very minor character flaws.

Kenneth sees Barbara as a little domineering at times. On this point, she considers herself simply self-reliant and independent, as any 21st century woman would be. One Saturday evening, when they went to dinner, Kenneth displayed his chivalry and held out the chair for Barbara. She told him, ever so gently, "Can I just sit down by myself? Yes, thank you very much."

Barbara saw Kenneth as unthinking at times. On this point, he considers himself simply ... unthinking. When it was time to (finally) meet her parents, Barbara planned a nice weekend lunch at a casual restaurant. Kenneth dressed in clean blue jeans and a T-shirt that read: BEAT NOTRE DAME.

Despite these small differences, soon Barbara learned to accept Kenneth's social graces and Kenneth learned that his future father-in-law went to Notre Dame.

He shredded the shirt.

And finally, Kenneth chose the most romantic setting that he could think of—near the potato chips at the supermarket—to ask Barbara to marry him. With tears welling in her blue eyes, she said "Can I just sit down by myself? Yes, thank you very much."[13]

Summary of the first chapter

Turbulent air screws up images. Horace Babcock came up with the idea for adaptive optics, a way to compensate for the turbulence and unscrew the images. Military developments from the 1970s to the early 1990s produced the technology necessary to make an adaptive optics system.

Barbara and Kenneth are dating.

---

[13] Which we all interpret as a "Yes!"

Chapter 2

The Atmosphere Has Gas

Every substance such as glass or plastic that is transparent to light has a characteristic called the *index of refraction*. This *index* is simply a number, from 1 on up, that tells us how that substance slows down light. A vacuum[1] is where there is no substance at all, sort of like modern political debates. The index of refraction of a vacuum is simply 1.00. In a vacuum, light travels at a very high speed. Light of any color (red, blue, pink) and even the light that we can't see[2] all travel at 186,000 miles per *second*. A beam of light can go around the entire planet Earth seven times every second. Santa Claus is the only comparable entity.

Curiously, anything other than nothing is called matter. I have sometimes called it "stuff," but was met with severe criticism from my English teachers. When light hits something made up of stuff (matter), it slows down. If light goes directly into matter, it keeps going in a straight line and slows down, so when it comes out the other side, it has only lost a little time. Compared with some light

---

[1] *Vacuum* is a weird word with double "u" and not a "w." Just say this out loud.

[2] Infrared heat, microwaves, ultraviolet sunburn rays, Golden Oldies radio waves, that sort of thing.

traveling along side, without going through the stuff, the first beam is said to be *delayed*. Much of this was explained by great minds hundreds of years ago.

What's really neat is when the light hits the matter, but not directly head-on. If it comes into contact (figuratively speaking) with the surface of the matter at an angle, the slowing-down property (the index of refraction) makes the light beam appear to bend or change direction when it enters the stuff. This property of matter, being able to bend light, is almost always useful. We can make prisms and see the wonderful colors of the rainbow without having to stand outside in the rain. We can make lenses of various shapes to focus light onto digital cameras and take really cool pictures. We can use other lenses to help us see clearly when our eyes are somehow not able to do it for themselves.

The speed of light through stuff

Numbers are very useful for describing the stuff. Remember the vacuum with the index of refraction of unity (that's 1)? Water, a very common substance that is colorless, odorless, and tasteless, has an index of refraction equal to 1.33.

[WARNING! WARNING! EQUATION APPROACHING!]

The speed (v) of light in a substance is

$$v = \frac{c}{n},$$

where c is the speed of light in a vacuum and n is the index of refraction of the substance. Water slows down the speed of light by 25%. We simply divide the speed of light c by the index of refraction n to find the speed of light in the stuff. 186,000 ÷ 1.33 is 140,000, a reduction in speed by 25%. Water, with its index of refraction of 1.33, bends a ray of light.

Try to read this book by looking at the pages through a glass of water. First, get yourself a clear glass, not the one with your favorite Sesame Street characters blocking the view. Fill it half way with water. It is now half empty.[3] Put this book behind it and read the text through the water. It looks distorted, right? Now add a few clear ice cubes. Take a sip of the ice water. Read the book. It looks even more distorted. Now add a shot of scotch. Take a few sips. Now you look distorted.

Other materials have other indices of refraction.[4] Glasses and plastics can be made with many different indices. An index of 1.5 is a popular one. Glass that has an index of 1.5 can slow down the light by 33%. Some glasses can have an index as high as 2. Crystal glass has an index of 2 and bends light in a special way to make chandeliers and wine decanters really nice to look at. If you have a big budget for scientific instrumentation, buy some Waterford Crystal and just call them high refractive index acousto-optic resonant cavities. Your mother, and surely Barbara's mother, would like it as a gift.

Other materials that we all know and love and cannot afford have really high indices. Sapphire is 1.77 and diamond is 2.4. Because the index is so high, a lot of the light gets into the diamond but can't get out right away, so it bounces around inside for a bit. Because of this phenomenon called *total internal reflection*, the specially cut angles to the diamond make it sparkle and glisten. Barbara, the new fiancée understands total internal reflection.

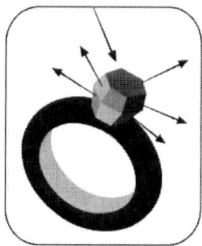

An expensive way to experience refraction.

---

[3] Optimists would say that it is half *full*, but they are just being optimistic.

[4] *Indices* is the correct way to write the plural *indexes*.

How did we get on the subject of diamonds and scotch on the rocks? The air, right, we were talking about the air. The index of refraction of air is a nice unround number somewhere near 1.0003, which means that air slows down the light by about 3/100 of a percent. And, as in the case of glass or water, it bends it a little bit, too.

Air is made up of various gases,[5] the most common being nitrogen (78%), with oxygen (21%) following at a distant second. The last remaining 1% is a mixture of carbon dioxide, necessary for all plant life; water vapor, the most absorbing of all the greenhouse gases but necessary for rain; and rare gases like neon, krypton, and argon, necessary for welding my car back together when the idiot behind me was on his cell phone and ignored the stop sign while he was stuffing a French fry into his face and ...

Sorry.

Let's look again at the index for air, 1.0003. It is really close to unity or exactly 1.0000000. But thanks to being close enough to the Sun to have a nice warm day at the beach, the air is moving around. When it does that, things like the air pressure and temperature vary. Air pressure is measured in millibars, which is a thousandth of a bar.[6] Temperature is measured in Fahrenheit or Celsius, depending on where on the planet you were born. So, it turns out that the index of refraction of the air depends on the precise values of temperature and pressure. It is much more sensitive to temperature than it is to pressure. A slight change of temperature, by about 1° Celsius, will change the index of refraction by 0.000001 or one millionth. This nifty little number means that a 1° C variation in a pocket of air changes the speed of light by 50 miles per second out of 186,000. This may not seem like much, but to a beam of light, it is huge.

---

[5] Being similar to your town council.
[6] ... which is usually quite busy on a Friday night.

If the air were a homogeneous mass with the same temperature, pressure, and density everywhere, we would not have a problem. As we know, the air is generally warmer near the surface at sea level and it gets colder as we climb higher. Ski resorts have never done well in the Bahamas. As the Sun, or any other warm source, heats the air, it expands. When it does, the density goes down slightly. This bit of warmer air can float over the cooler air. Hot air balloons rise on this principle alone. The lower density also has an optical effect. There is a small, but noticeable reduction in the index of refraction. A beam of light going through the warmer air will not be delayed as long as the one in the cooler air is delayed.

All this would be no problem if all we wanted to do was to wait for the little bitty beam of light to get somewhere and we didn't care when. It is a big problem if we need to collect a whole bunch of little bitty beams all at the same time.

The wedding day

Think of the big Barbara and Kenneth wedding coming up. The bride and groom have planned for this for years, or maybe weeks, or maybe about 20 minutes in Las Vegas. The point is that everything must come together at the same time or else it won't work. The bride and groom must both be at the same place and at the same time.[7] The minister/rabbi/Elvis impersonator must be there, too. The witnesses must be there and be able to sign their names or aliases. Nobody can be late or the whole deal is off. In many cases, bridesmaids and groomsmen must be around to embarrass themselves later at the reception. Parents, uncles, aunts, various stepchildren, and former girlfriends may also be in attendance. All this has to happen in an organized fashion so the marriage can take place, and the newlyweds can move on to the honeymoon, the wedded bliss, or divorce court.

---

[7] This is usually standard behavior at a wedding. I'm not quite sure why.

Beams of light are like that too, except for the former girlfriends. For the sake of this ridiculous analogy, let's assume that the wedding photographer was there to take pictures. He arranges the bride and groom into a highly fixed pose that is to resemble a candid I-didn't-know-you-were-taking-my-picture pose. The sunlight from the window streams onto the blushing bride and thirsty groom. The photographer raises his camera and CLICK, a memory that will last a lifetime!

These thirteen people at the Tyson-Castro wedding, plus the photographer, had to arrive precisely at this place, at this time in 2008. All 13 subjects had to execute a reasonable smile, including my brother (far right) who was paying for the whole thing. (Photo courtesy of Bella Pictures®.)

Now for some science: The picture was formed when zillions (technical jargon for "a lot") of light rays from all the different parts of the bride and groom scene somehow slipped through the camera lens and formed an image on the camera's electronic sensor array (the megapixel thing that costs so much). For the image to closely resemble the bride and groom, a lot of things had to happen.

The light rays from the real bride and groom have to be going in the direction of the camera lens. (A good photographer generally knows how to point the camera at the subject.) Next, the lens and other optical gizmos in the camera must focus the light almost perfectly to get an image of Barbara and Kenneth that is not distorted and makes them look like they did last night at the rehearsal dinner. And finally, the camera gizmos have to store away that image.[8]

In most cases of wedding photography, the bride, groom, photographer, and assorted in-laws are in the same room or garden and there are few problems with air being warmer in one part of the room than the other.

Untwinkling the stars

Now, think of some celestial wedding. Instead of two *people* who are *stars* for the day, imagine two *stars* that really are stars, probably for a lot longer than one day. The stars are a trillion miles away. Professor Leslie B. Smythe, the observational astronomer, has to take a picture of them. The light from the pair of stars is pretty faint. So faint, in fact, that you not only have to open the camera shutter[9] for a long time, but you also have to worry about the apparent motion of the object up there during the time the shutter is open. The Earth is rotating at 1000 miles per hour. Imagine trying to take a picture of Barbara and Kenneth cutting the cake with the light of one candle somewhere in the next county. Romantic, but stupid. The camera grabs light coming from all different directions that get overlaid on the image and you get a big blur, similar to the blur you get when trying to remember the wedding day.

Prof. Smythe, the astronomer, has even more trouble to worry about. The light from the bride and groom stars travels quite nicely for a trillion miles. And then, just as he opens the shutter of his multimillion-dollar megapixel giant camera, the light beams from

---

[8] To, you know, like, really, irritate the former girlfriends.
[9] Door for the light to go through when it wants to go into the camera.

the stars hit the atmosphere. The index of refraction is on average 1.0003. But some parts of the air are a bit warmer than others and each degree changes the index by 0.000001 and speeds up the light by 50 miles per second. Then again, some parts of the air are a bit cooler than others and each degree changes the index by 0.000001 and slows down the light by 50 miles per second. What a mess!

All the zillions of light beams have to get there at the same time! If all are going to arrive at different times with different time delays, the marriage is annulled before it gets started. The picture that Prof. Smythe takes looks not at all like the bride and groom stars, but rather like Uncle Fred after the reception.

Prof. Smythe can't take the picture for a shorter period of time, because the stars are too faint. Over the long exposure, the atmosphere keeps changing, randomly no less. As we defined earlier, the atmosphere screws it up.

Fortunately, we do understand a lot about what is going on with the atmosphere. We can't predict the weather for an hour from now, but we can get an idea of why the atmosphere's index of refraction changes. This has been explained with some nifty theories and calculations by Kolmogorov and Tatarski. It doesn't matter who they are, although I admire their work. I'll bet that you're glad I didn't name our bride and groom Kolmogorov and Tatarski.

Seemingly random thoughts about statistics

The changes of temperature in the atmosphere that are screwing up the light rays are random. We can't predict exactly what they will be at any one instant, but we do know something about them. This allows us to make some general predictions, like averages. This is not as strange as it seems.

One-half of a pair of dice is called a die.[10] It is a cube and on each of its six sides is a number from one to six. If we roll the die by tossing it from our hand along a surface, there is an equal probability that any of the sides will end up on top. We can't predict that it will be a three or a five when we throw it, or any other particular number, but we can predict that if we throw it many, many, many times, the average of all the numbers showing up will be 3.5 and there will have been roughly as many ones as twos and twos as threes, and so forth.[11]

A perfectly worthless purpose for a pair of dice.

If I throw the pair of dice in a nicely lit casino, where people are walking by and doing everything they can to distract me, and cocktail waitresses are offering free drinks, expecting only a nice tip, and ... I'm off-message here. Back on track ... Say you add the sum of the two dice each time, over many throws; the average will be seven. The scientific term for the seven is *craps*.

---

[10] It then follows that one half of a pair of grains of rice is called a rie and one half a pair of lice is called a lie. A person who is only being half nice is called a nie.

[11] What is neat about this little example is that the average 3.5 is not even any possible result of a single throw.

We understand very little regarding how to predict an individual occurrence (a *realization*), yet we understand what the average will be. And the word "craps" has a nice ring to it, doesn't it? With dice and many other random happenings, we can often figure out how far away an individual attempt may be from the real average (of many attempts). This interesting statistical quantity is called the *variance*.

A statistician at work. Notice that he has only three fingers and a thumb on his left hand. Presumably, he is fluent in base-eight mathematics.

Well, so it goes with the turbulence in the atmosphere. We can calculate and then measure averages and variances all over the place. The random changes in index of refraction in the air happen both spatially and temporally. If we move through the space in the air from point to point, the index will be random, with known averages and variances. This is *spatial* randomness. There is also *temporal* randomness.[12] *Temporal* here refers to *time*. If we stay at one point in the air space, the index will change with time, also in a random fashion, with known averages and variances.

---

[12] This happens if we move from one prehistoric temple to another.

So, by knowing these averages and variances, commonly called *knowing the statistics*, we can predict what the atmosphere will do over a wide range of distances and over a long time. As we think of building an adaptive optics system to fix this nonsensical random mess, the problem for us is: What are the spatial and temporal statistics?

Both statistics exist at the same *time* all over the *space* of the atmosphere. The space already has three dimensions: height, width, and breadth. Thus, the atmosphere throws us a four-dimensional problem if we include time. We don't have to be arbitrary with the spatial dimensions. When worrying about optical turbulence, we are dealing with the surface of the Earth up to an altitude of about 40 kilometers (120,000 feet or 25 miles). There is actually some air above 40 kilometers, but there is so little and the molecules of nitrogen and oxygen are so far apart, that they have little effect on a beam of light.[13]

For simplicity, we'll split up the study of the turbulence into three pieces. The first piece combines two of the dimensions. We can take the north-south and east-west directions and just create one flat surface (a mathematical plane). Think of a street map except it has numbers that represent the index of refraction at each place rather than the locations of schools, houses, or bookstores.

The second piece is altitude, so think of having the street map moving higher and higher and changing at each altitude. The third piece is time, measured in something like milliseconds. Because the atmosphere is heated by the Sun at all altitudes, we end up with big blobs of warm air. A precise scientific size of a big blob is somewhere around 100 meters, maybe more, maybe less. Luckily, the index of refraction throughout the big blob is the same and won't affect our beam of light much at all.

---

[13] Or on a wedding photograph.

Big blobs and little blobs

Now, when we add natural phenomena like wind and the spinning Earth, we get the big blobs to break up into smaller blobs. And this keeps happening until the blobs get so small that they can't break up anymore. A precise scientific size of a little blob is somewhere about 1 millimeter, more or less.[14] When the atmosphere is near the surface, there are a lot of natural and manmade things to further break up the blobs. Wind traveling over mountains and around buildings will do it. Heat, not directly from the Sun, but from buildings, cars, and Buffalo wings will do it. Millions of random moving blobs—it sounds like the title of an Oliver Stone movie about the latest Palm Beach weight loss program.[15]

The light going through one small blob will not add up properly to the light going through another small blob when they get to the telescope to form an image. One is delayed a little bit longer than the other. Consider that for the image to form, we need zillions of these rays and now they are all experiencing their own little random delays as they go through the blobs.

The starlight has to travel from the top of the atmosphere all the way down to the surface of the Earth. Forty kilometers is a long way, but it only takes a beam of light 1 millisecond to do it. When it does, it goes through all these blobs, or eddies, and they are all different sizes and shapes and they keep changing and moving and they just never take a break. Then the rays eventually are collected by the neat optics of the telescope. Without adaptive optics, the image is all distorted.[16]

---

[14] The scientific name for the big blob is the *outer scale*. I prefer Jabba the Hutt. The name for the little blob is *inner scale*. I think of Yoda. The scientific name for blobs is *turbulent eddies*. I think of Eddie Miller who bullied me in third grade and is now thankfully serving 10 to 20.

[15] It is not a movie. It is real and it is coming to a telescope near you.

[16] Up to a point.

Air blobs (turbulent eddies). Turbulent Eddie Miller is in the lower left.

A light ray going through the turbulent eddies.

It is not completely screwed up. It turns out that there are some statistical averages of all these eddies that we know about. This turbulent random atmosphere can be averaged in a way that we can add all the eddies at all the altitudes and determine the size of a telescope that will allow the light to form a near-perfect image. In the mid-1960s, David Fried[17] did this important calculation. This distance now has various names, such as coherence length, scale size, seeing-cell size, Fried's parameter, or Fried's coherence length, and usually uses the mathematical symbol r_0, so it is often just referred to as "Ar - naught." These terms all mean the same thing: the limiting size of a telescope that will produce a good image in spite of the turbulence in the atmosphere.

How big would this telescope be? It depends on two things: the wavelength or color of the light and how bad (or good) the natural seeing conditions are. For longer (redder) wavelengths of light, the r_0 is bigger and for less turbulent atmospheric conditions (a cold clear night), it is bigger. We want bigger, so that we can use big telescopes to see fine detail. Unfortunately, we can't always use longer wavelengths or wait for a once-in-a-lifetime night for looking at the nifty things in the sky. So, we just have to deal with it.

It's unfortunate, but for the middle of the visible light spectrum where our eyes work the best, and in many places where we want to put our telescope, the coherence length is only 10 centimeters (4 inches). The biggest telescope input lens diameter that we can use, without the atmosphere screwing it up, is only four inches! Regular bird-watching binoculars, used by irregular birdwatchers, are about the limit.

Have you ever used a pair of binoculars[18] to look at something far away? If your answer is YES and the *something* was Barbara's kid

---

[17] D. L. Fried, "Limiting resolution looking down through the atmosphere," *J. Opt. Soc. Am.* **56**, 1380, (1966).

[18] *Bi* as in two of them, *ocular* as in optical thingy; the *n* is there to make it pronounceable. The *pair* of binoculars is redundant, like a *pair* of pants.

sister or Kenneth's older brother, then that is okay. If your answer is YES and the *something* was perfectly innocent, then that is okay. If the answer is NO because you have never used binoculars, then stop reading. Stop reading now. You do not have the qualifications to continue reading this book. Please close it gently and donate it to a library in the Bahamas. I will personally deliver the book to the library. Send the book to me with a check for the travel expenses and I will take care of it for you with kindest regards.

If you are still reading, then I applaud you.

Fried's coherence length (r_0) essentially limits the size of the telescope. With a larger telescope, we can collect more light, but the atmosphere reduces the size of the detail that we can see. If we want to make use of the big, big telescopes, we have to do something extraordinary. This is a job for Superman![19]

**Low resolution. The telescope is too small
and you can't distinguish details.**

---

[19] The atmosphere contains krypton. Superman is not just some powerful fantasy creature. He wears his underwear outside his clothing.

A larger telescope can distinguish the details.

Summary of the second chapter

The atmosphere is made up of gases that get heated by the Sun and form blobs that move around. The motions are random, but some of the statistics of the motion of the air are understood. The index of refraction is a number that describes how much the air slows down a beam of light. The physical size of the air blobs is called Fried's coherence length.[20]

Barbara and Kenneth get married and soon begin to squabble.

---

[20] And other names.

Chapter 3

Adaptive Optics Systems
and
Some Cool Things About Light Beams

A job for Superman? He is way too busy fighting for truth, justice, and the American Way.[1] It's a job for an adaptive optics system: an eclectic combination of mechanical gizmos, optical widgets, and electronic thingamabobs. When they all work together, they have at least a fighting chance of fixing all that mess that the atmosphere made.

To do this nearly miraculous thing, the adaptive optics system must first figure out what the atmosphere is doing. Then it must figure out how to fix it. And then finally, it has to actually go ahead and fix it.

---

[1] In his spare time he is using his x-ray vision to scan luggage at O'Hare Airport.

How a relationship is like adaptive optics

This is much like the work of Dr. Mannheart, the marriage counselor for Barbara and Kenneth. Note: The honeymoon is over. In fact, it was over as soon as Kenneth left the cap off of the tube of toothpaste. Day three, to be precise.

The counselor talks with both warring parties (diplomatically described as *the parties* in the minor domestic dispute). First, Dr. Mannheart figures out what the problem is. Then he proposes a solution. Finally, it is up to Barbara and Kenneth to actually carry out the solution, presumably without too much further damage to the furniture. Like an adaptive optics system, everything must work together. If the counselor misdiagnoses the problem, the solution probably won't work. If the diagnosis is right and the solution should fix it, but Barbara and Kenneth don't stop screaming at each other long enough to do what the counselor suggests, it won't work either. Finally, if a new problem crops up, the whole process must be done all over with a different solution and really good behavior by the newlywed couple.

That's the way it is with an adaptive optics system. The marriage counselor role is played by a subsystem called a *wavefront sensor*. The proposed solution role is played by a subsystem called the *reconstructor*. And finally, the shut-up-and-do-the-right-thing role is played by a subsystem called a *deformable mirror*.

Each one has its own special function. Each one must work together with the others. When it comes to the doggone atmospheric turbulence, each one has to solve a brand new problem 30 times every second.

I used the term *subsystem*. This term and many like it come out of engineering complicated things and then keeping track of them. If an adaptive optics system is composed of some pieces, we call the pieces *subsystems*. The subsystems are made from *assemblies*. The assemblies are made from *subassemblies*. The subassemblies are made from *components*. And components are made from subcomponents.[2] It happens to go in the other direction, too. For example, an adaptive optics system can be part of a larger device called a beam control system, which should really be called a *supersystem*, but very few people actually use that term.

Anyway, the adaptive optics system will always have these three subsystems in one form or another: a wavefront sensor, a reconstructor, and a deformable mirror. Their purpose is to work together to fix the problems they encounter. In most cases, what they fix is the *phase* of the beam of light, in a process called *phase conjugation*.

Phase conjugation is not a medical condition. It simply means "opposite phase." That's what an adaptive optics system does. The wavefront sensor measures the phase of the light coming through the blobs in the atmosphere. The reconstructor calculates the opposite phase and sends little tiny messengers dressed in green leotards to the deformable mirror. The deformable mirror does what it is told and shapes (deforms) the mirror surface into the opposite phase that it was told (as best that it can). By doing this, the messed-up phase of the beam is un-messed-up, by the process of phase conjugation. Now all the little delays from the turbulence are ironed out and a really nice clear image is formed.

It doesn't phase me anymore

It is time to describe what *phase* actually is. (Heck, it's about time!) Yes, really, it is *about time*. It is related to the time delay of one beam of light to another.

A beam of light is also a wave of electromagnetic energy. Waves— from water waves to waves along a violin string to shock waves and sound waves—have a few important characteristics. They have strength or *amplitude*. In a sound wave, that's the loudness. In a

---

[2] This is all very logical to a nerd with a pocket protector and too much time on his hands.

light wave, that's the brightness.[3] In a water wave, it is the height of the wave above the average sea level. Surfers look for waves with lots of amplitude. In addition to amplitude, waves have *frequency*. In a sound wave, that's the pitch or tone. In a light wave, frequency is the color. On a water wave, frequency is how fast the next crest comes in.

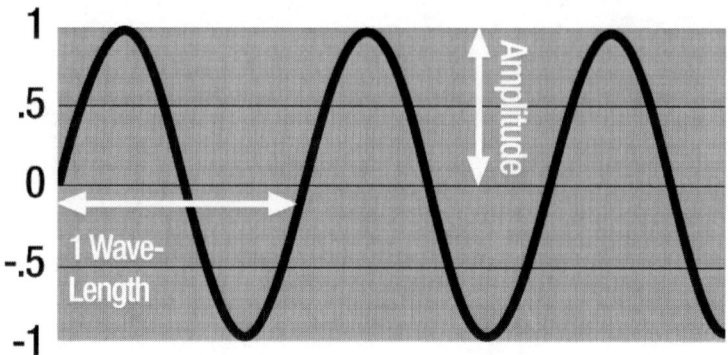

Amplitude and wavelength are characteristics of a wave.

Related to the frequency is the *wavelength*. Waves, by their very nature, wave back and forth. They are constantly repeating themselves at a constant frequency. Unless someone or something gets in their way, they'll go on forever, or a very long time, anyway. One wavelength is how far the wave travels before it repeats itself. How far it travels, of course, depends on how fast it is going.

[WARNING! WARNING! EQUATION APPROACHING!]

The relationship

speed = frequency × wavelength   (× means multiply by ...)

applies to all waves: sound waves, wind-driven water waves, waves along a rope, and all of the other waves, including light waves. For

---

[3] Yes, I know that brightness is really the *square* of the amplitude, but does that really matter to Barbara and Kenneth at this point?

example, sound waves in air travel about 770 miles per hour (343 meters per second ... almost four football fields in one second). The frequency of the "A" tone on the musical scale is 440 Hertz. (1 Hertz is the same as 1 cycle per second.) So the "A" tone repeats 440 times every second. From our simple formula, the wavelength of the "A" tone is 0.78 meters or about 30 inches. Thus, a musical wind instrument has a length of 30 inches to play an "A." To play other notes, every modern wind instrument has a way to change its length, either with longer paths, as does a trumpet or trombone, or by forcing the air to turn around at a different place with holes or keys, as does a flute or clarinet.

You have to be careful to do this calculation for stringed instruments because the speed in the formula is not the speed of sound in air; it is the speed of the wave along the string itself, which depends on thickness, weight, and tension.[4]

That finally brings us to *phase*.

Imagine Barbara and Kenneth going to a symphony orchestra concert. They want to forget their little differences and enjoy each other again. I know that I am getting very abstract here, but it's to make a point. They drove all the way into the city, waited endlessly in traffic, parked their car for something like the price of an ounce of gold, and gave their tickets to the doorman (actually, it was the doorperson). The concert is to start at 8 p.m. At about 8:15, the members of the orchestra are all seated and in walks a guy with a violin and everybody applauds.[5] He looks at the oboe player and nods. The oboe player *finally* plays a note ... one note, probably an "A" at 440 Hertz. Now all the musicians pick up their instruments and make what seem like random sounds. They are tuning their instruments. A stringed instrument is usually tuned by adjusting the tension, which changes the speed of the wave.

---

[4] My most recent New Year's resolution was to remove excess thickness, weight, and tension.

[5] No one has even played a note yet! What are they applauding? The fact that the guy knew where to find his tuxedo!?

When it sounds just right—and most trained musicians are *very good* at knowing when this happens—the speed of the wave is such that the frequency is just right. Let's follow the very front of the wave as it travels along the string. It will travel along the string to the end and then, when it reaches the end where it is fastened, the wave will reflect back and travel toward the other end. There it reflects again, back and forth, back and forth. The travel time is perfectly suited to doing this 440 times per second. The string will be vibrating at just the right frequency.

It's very important that the front of the wave going one way meets the front of the wave coming the other way. All along the string, one wave is moving up the string when another is moving down the string. If both waves are in the "left" position or the "right" position when they meet, everybody is happy. If one wave is moving "left" when it meets the other moving "right," they are said to be out of phase. The wave won't sound very good. If you have never played a violin yourself, borrow one from your boss and try to play it.[6]

Phase is measured by how far the front of the wave lags behind another (reference) wave. Because waves keep repeating themselves, because waves keep repeating themselves, because waves keep repeating themselves, they go around like riding a bicycle in a circle. You can keep going and never stop, but you end up at the same place.

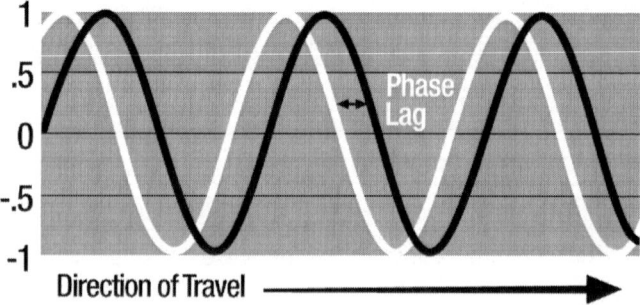

When two waves (even with the same amplitude and wavelength) start at different times, there is a phase lag between them.

---

[6] You will be not only out of phase, but probably out of a job, too.

Phase is measured as if it is traveling on a circle. The phase of the wave can be 10 degrees behind the reference. It can be 90 degrees (one-quarter of a wavelength) behind. It can be exactly one-half a wavelength behind (180 degrees out of phase). Or, it can be so far out of phase (360 degrees) that it is back *in* phase. Remember, the phase doesn't mean anything until it is compared to another wave.

The same thing happens with light. It can be in phase or out of phase with another beam of light. When it is in phase, we have what is called *constructive interference*. The two light beams, because of their wave nature, are helping one another. When they are out of phase with each other, we get *destructive interference*. The beams actually cancel each other out. We can't even see any light when this happens.

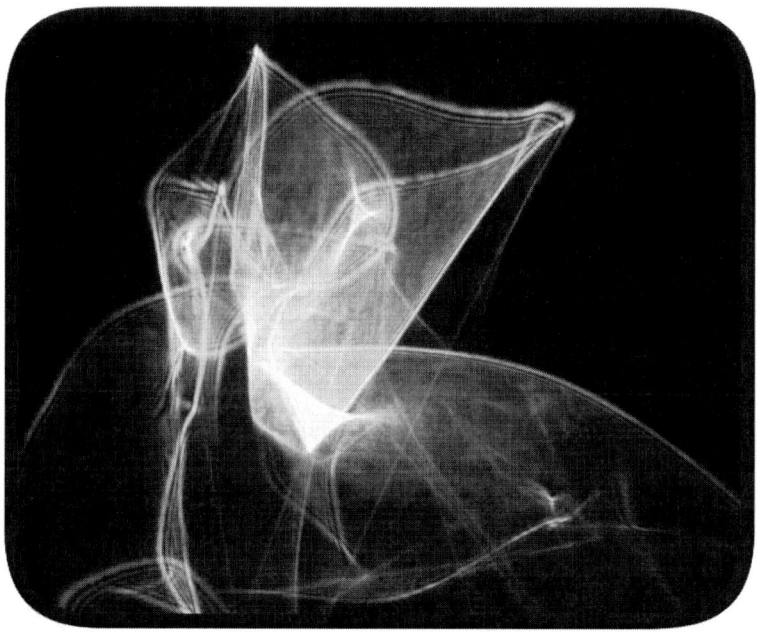

Laser beams can have very complicated and seemingly artistic interference patterns, but they don't make their creators depressed or egotistical. Usually. (Photo courtesy of Shutterstock.)

Two beams can pass right through each other and go on their merry way. Try crossing two flashlight beams. They don't even know that the other one is there.[7]

It is now time to build the subsystems and components of the adaptive optics system. Not literally, of course. Let's just talk about the system. We need to measure the time delays of the incoming light. We do this by measuring the phase of the light. This is not always simple. We can't go to a home improvement store and look in the power tools section and find an optical phase meter doohickey. We need to measure the phase by comparing it to another reference beam. We can figure out the phase if we find out where the front of the wave is. Our astronomer, Leslie B. Smythe needs a *wavefront sensor.*

Summary of the third chapter

Beams of light have intensity and phase. The atmosphere slows down one beam with respect to another beam and they interfere with each other, blurring the image. An adaptive optics system consists of a wavefront sensor, a reconstructor, and a deformable mirror, all working together.

Barbara and Kenneth get the advice of Dr. Mannheart, a marriage counselor, and begin to address their difficulties.

---

[7] Do not try this with *Star Wars* light sabers.

Chapter 4

Clever Wavefront Sensors

The wavefront or phase of a beam is only meaningful when we compare it to a reference. Prof. Leslie B. Smythe needs a reference beam to measure how the atmosphere has distorted the phase. Being an astronomer, he decides to use one of the stars in the sky. This is very clever because (1) there are a lot of stars in the sky, and (2) they are very far away. Except for the closest star, our Sun, they are all so far away that they only appear as points of light. They don't seem to have any apparent size. The light seems to leave the same point on the star at the same time. This phenomenon, called *coherence*, is going to be a great benefit for our wavefront sensor. The light travels out from the star equally in all directions. The wavefront of the light from this star, as it moves outward, is a giant sphere. Because we know precisely what it is, we can use this light as a reference.

Did you ever think that the world was flat? You probably did before you became a highly educated person and realized that, even though you know the truth now, it seems flat. (At least it does on the surface.) So, perception is reality? No, it's just that the area around you, on this large sphere Earth, is so small that when it is compared to the 4000 mile distance to the center of the sphere, it appears flat.

The stars are really far away. The closest star is 20 thousand billion miles away. By the time the light from this star reaches Prof. Smythe's little telescope, the wavefront is totally indistinguishable from flat. This makes a really great reference beam. From this point forward, we can refer to the reference star as a wavefront *beacon*. The wavefront is a flat plane as it enters the Earth's atmosphere. After it goes through all the blobs of air, the wavefront is no longer flat. It is like flattening out a crumpled newspaper, just with really tiny crumples. The crumples contain information only about the atmosphere. Now all we have to do is compare the crumples to another flat reference and we have a measurement of the phase distortions of the turbulent atmosphere.

Let's picture, with our minds, a wavefront.[1] I will do it in two dimensions, instead of three. Our sphere becomes just a circle. Imagine throwing a pebble into a still pond. The little waves make circles coming from the point where the pebble hit. I hate this analogy! It's too Henry David Thoreau.[2]

Ripples in Walden Pond.

---

[1] If you can picture one with your pancreas, please go ahead and avoid telling me the details.
[2] Nobody ever threw a pebble into a pond. It's an urban myth.

Try this. Imagine throwing a 24-pound frozen turkey into a swimming pool. That's better. It happens all the time. The big waves make circles coming from the point where the turkey hit. The waves move outward. They have a front (in the direction that they are going) which is called the wave*front*.[3] Now imagine that the swimming pool has some children in it. (I like this part.) Throw in another frozen turkey, being sure not to directly hit one of the poor dears. The waves still move outward in circles until they are disturbed by the screaming children. The wavefront is no longer circular; it is disturbed, just like the wavefront of the light coming through the atmosphere from distant stars. The wave without the children can be used as the reference wave, while the disturbed wave with the terrified six year olds is the wavefront we want to measure.

The right way to make ripples and bring joy to little children.

But wait a minute. Wait just one minute! We need to know a precise measurement of the wavefront in order to send the

---

[3] I suppose they have a back too, but no one ever called them a wave*back*.

information (about the problem) to the reconstructor (so it can solve the problem). How do we actually measure the wavefront? We need an electro-optical[4] wavefront sensor. Some clever scientists worked on this, and by the 1970s, there were a few operating wavefront sensors. To this day, scientists and engineers are improving them and inventing new ones.

Shearing interferometer

The shearing interferometer was not invented by anyone named Shearing. It refers to the way the incoming wavefront is compared to a reference wavefront at the pupil of the optical system. The reference beam is not a separate flat wavefront beam; it is just a replica of the crumpled incoming wavefront, with a trick ending, just like in a Hitchcock movie. The trick is taking the reference version and shifting it slightly in one direction. The shift is called the *shear*, hence "shearing" interferometer. When the crumpled wavefront is added to the crumpled (but shifted) wavefront, the interference (constructive, destructive, whatever) forms a pattern of light regions and dark regions that can be used to calculate the crumpled wavefront.

This type of wavefront sensor is not used much these days, because it has a lot of moving parts to make the shear and detect the faint light, thereby needing a lot of tender loving care to keep it going. I mention it because it was the type of sensor used on the first space surveillance adaptive optics system on Mt. Haleakala in Hawaii, circa 1982.

Shack-Hartmann wavefront sensor

The Shack-Hartmann sensor was invented by Shack and Hartmann, although not working together. Johannes Hartmann invented a

---

[4] Scientists like to abbreviate. *Electro-optical* just means something is both electrical and optical. We often abbreviate it further to simply E-O. Then we find other things with the same letters, such as Enemy Operation, Elementary Ornithology, or Extra Olives to confuse you.

wavefront test at the beginning of the 20th century. The test is fundamentally very simple. We want to see spots in front of our eyes without actually getting hit on the head.

The beam of light being tested (also at the pupil of the optical system) shines on an opaque screen, where the light normally doesn't go through. An array of tiny holes is drilled into the screen so that light can get through at the holes. Behind the screen is an observation screen where the little spots of light hit. If the spots of light hit exactly behind the center of the hole, the light must have come through the hole perfectly straight. If the spots of light are not directly behind the hole, then the beam of light is tilted, at least at the entrance to the hole. The positions of all the so-called *Hartmann spots* create a "map" of the wavefront by measuring the little tilted beams at each of the holes. The Hartmann test has been used for a long time to test large optics.[5]

The problem with using the Hartmann test is that most of the light doesn't go through the tiny holes. It is lost. And for astronomy, there is very little light to throw away. If the holes are made bigger to let more light come through, the spots are larger and the precise position cannot be easily determined. Roland Shack, with the assistance of Ben Platt, of the University of Arizona, had a solution to this dilemma.[6] Instead of an opaque sheet with holes, Shack proposed no holes at all, but an array of lenses with each one forming a small focused spot on the observation screen. This way, all the light is used, and the spots are still small. Because the array of lenses is just a bunch of small lenses, they are called lenslets. Each little lenslet is an aperture (opening) in its own right, so each one represents part of the whole aperture. Each lenslet forms a *sub*aperture.

---

[5] *Large* here is defined as bigger than a breadbox, although no one remembers ever actually having seen a breadbox and knows how big they are.

[6] R. B. Shack, B. C. Platt "Production and use of a lenticular Hartmann screen," *J. Opt. Soc. Am.* **61**, 656 (1971).

And then to make matters worse, in the world of adaptive optics, the "Shack-Hartmann sensor" is often abbreviated to "S-H sensor" or sometimes called the "Hartmann-Shack sensor" or worse yet, called simply the "Hartmann sensor." And to make matters really, really worse, some misspell it as "Hartman sensor."

If the beacon for the wavefront is a point source, like a faraway star, the position of the center of the little spots is an accurate measure of that little bitty wavefront tilt. But if the wavefront beacon is an object that can actually be resolved (seen in detail), like the planet Saturn, then the Hartmann spots are itty bitty images of the planet Saturn. The "center" is often very hard to define because a bright area of the image will fool the computer.

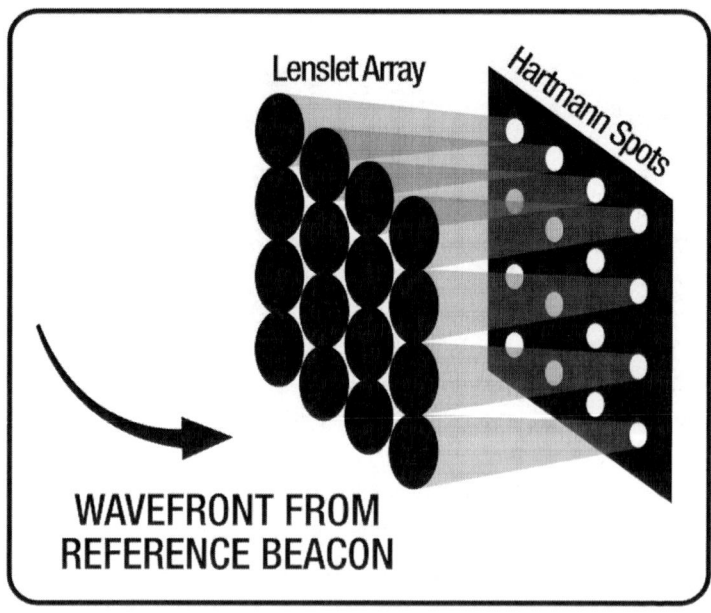

A Shack-Hartmann sensor has spots of light from a reference beacon.

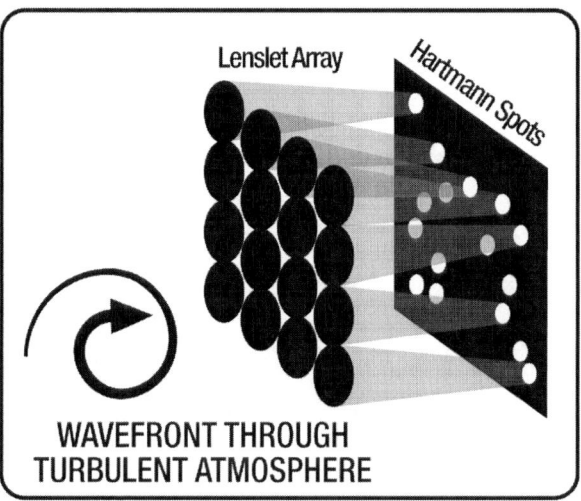

A Shack-Hartmann sensor has spots of light from the distorted wavefront after the light goes through atmospheric turbulence.

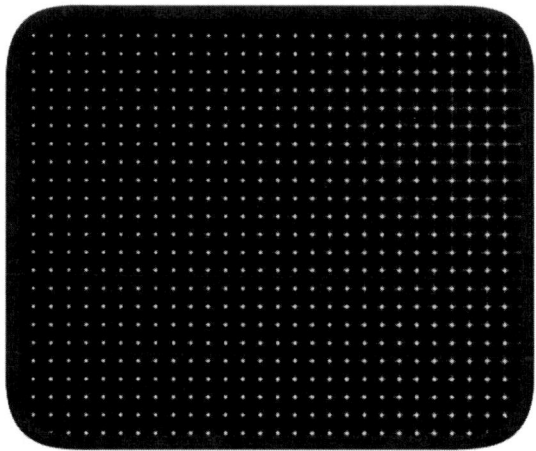

A real picture of Hartmann spots in a Shack-Hartmann wavefront sensor. Stare at this for three minutes. When you begin to feel sleepy, go to the bookstore and buy 35 more copies of this book. Thank you.

Roland Shack, Johannes Hartmann, and some more of their spots. (Photo of Hartmann painted portrait, courtesy of Michael Hartmann Collection.)

Curvature sensor

François Roddier at the University of Hawaii invented the curvature sensor.[7] Unlike the Shack-Hartmann sensor and the shearing interferometer, a curvature sensor works at the image, not at the pupil, and doesn't measure the localized wavefront tilt. Roddier determined that when a slightly out-of-focus image is subtracted from another slightly out-of-focus image, you don't have just blurred vision. When the exact amount of focus error is known from the subtracted images, you can calculate how much wavefront *curvature* is present at each point back in the pupil. Curvature is a measure of how curved the wavefront is at each point.[8] With a nifty computer, by knowing how curved it is everywhere, we can calculate a map of the wavefront that can be used just as the other wavefront sensors are used.

---

[7] F. Roddier, "Curvature sensing: a new concept in adaptive optics," *Appl. Opt.* **27**, 1223–1225 (1988).
[8] Technically speaking, curvature is the second derivative of the wavefront.

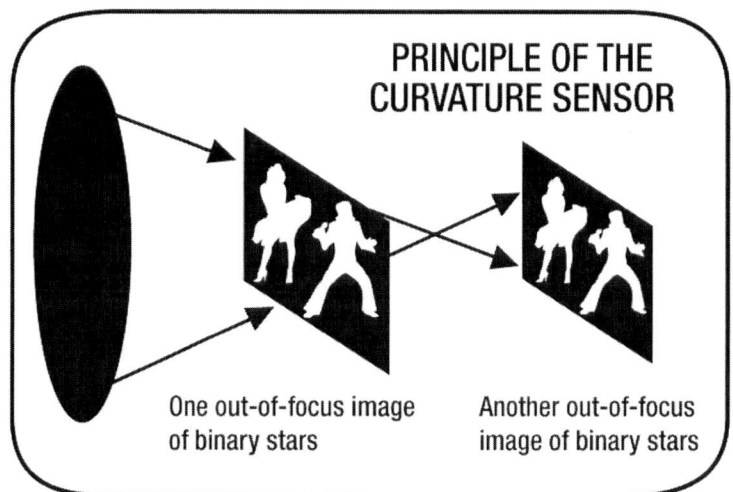

PRINCIPLE OF THE CURVATURE SENSOR

One out-of-focus image of binary stars

Another out-of-focus image of binary stars

Principle of the curvature sensor. Two precisely out-of-focus images are subtracted to reveal the secret hidden treasure.

There is one nice benefit to knowing the wavefront curvature. A special type of deformable mirror, called a bimorph,[9] can be driven electronically and the natural shape of the mirror surface is curved. This makes the reconstructor very simple, saving you money, time, and probably giving you better gas mileage.

Pyramid sensor

A pyramid sensor is useful when you are on a trip to Egypt and your tour guide misses the bus. You can still find Giza. Very few adaptive optics systems are found in Egypt, mostly because blowing sand just causes havoc with the optical coatings.

You may have heard of the popular myth that the pyramidal shape has magical powers. Something placed inside the pyramid can heal faster or the pyramid can mystically sharpen razor blades, or other

---

[9] Two morphs.

wonderful things can happen. Advertisements with testimonials from astonished beneficiaries of the magic of the pyramid tell us how we should have one for our very own. You can have one for $19.99 plus shipping and handling. I don't doubt that the commercialized pyramids have a benefit.[10]

On the scientific front, a pyramid, especially the four-sided kind, makes a perfect shape for dividing a beam of light into four replicas of itself. The pyramid wavefront sensor, invented by Roberto Raggazoni,[11] has a small glass pyramid at the position of the image. The image is divided into four versions and, by the magic of Fourier optics,[12] each one is sent back to a pupil.

Principle of the pyramid sensor. Sand and camels added for effect.

---

[10] The benefit is that the manufacturers are making money.

[11] R. Ragazzoni, "Pupil plane wavefront sensing with an oscillating prism," *J. of Mod. Optics* **43**, 189–193 (1996).

[12] It's not really magic. It turns out that light can propagate from a pupil to an image and then again to another pupil (a conjugate pupil) and then to another image and to another pupil and sort of go on forever and ever.

The light beams at the four pupils can be compared to each other and wavefront tilt can be found for each point in the pupil, which is pretty much all that the Shack-Hartmann sensor did. Except, the pyramid sensor has a much higher spatial resolution. It can sense more detail in the wavefront shape.

Thankfully, the pyramid of the pyramid wavefront sensor is much smaller than the great Cheops Pyramid. Prof. Smythe can use a pyramid that is about one centimeter or 0.5 inch. But, with the high-quality optical polishing and coating, it costs a bit more than $19.99 plus shipping and handling.

While each of these wavefront sensors works in a slightly different manner, they all have one purpose—to measure the crumples in the wavefront caused by some disturbance, such as vibrations or atmospheric turbulence.

Summary of the fourth chapter

A variety of wavefront sensors can measure the phase distortions of light when it goes through the atmosphere. The most common type is the Shack-Hartmann sensor.

During this chapter, Barbara and Kenneth went to the movies at the Super 49 Theaters Plaza. Barbara chose the movie.

Chapter 5

Laser Guide Stars, the Beacons in the Night

Without getting into such extremely boring topics as signal-to-noise ratio, detectivity, or noise-equivalent power, let's just say that wavefront sensors must have reasonably bright light in order to work. Shack-Hartmann and pyramid sensors need a point source of light to work best. But what if the thing that we are looking at is not bright enough? Remember, we must use some of the light for the wavefront sensor and then have enough light left over for the image to form, which is the reason for doing the whole thing in the first place.

Scattering is the problem

Objects in the sky that are scientifically interesting are called *science objects* (the things we want to look at). Very few of the science objects in the sky, like far away nebulae, star clusters, or galaxies, are bright enough to be a wavefront beacon. One solution is to find another nearby bright star that can be used as a beacon for our dim science object. This is a great idea, but it hits an early snag. You can't look at the sky in two different directions and find the same atmospheric turbulence. That 10-centimeter blob of air is really very small and when we imagine it high in the sky, the angle from one blob to the next is also very small.

I occasionally like to use big words. But, I won't just insert big words like pneumonoultramicroscopicsilicovolcanoconiosis without an explanation. Imagine the sky as one flat sheet high above us. Because it's flat, like a plane, we say it is *planatic*. If the turbulence were the same all over the sky, no matter which direction we looked, the atmosphere would be *isoplanatic*. But it isn't. We only have to look in another direction by about 4/10,000 of 1 degree to find a different turbulent wavefront. The atmosphere is an*isoplanatic*. The word anisoplanatic is not as long as pneumonoultramicroscopicsilicovolcanokoniosis, but it gives me an excuse to use the longest word in the English language[1] twice in a book that has nothing to do with a nasty volcanic-dust lung disease.

So, now we have reduced the problem to this: Our science object is not bright enough to serve double duty as a wavefront beacon and we can't look very far away from the science object and still have a good measure of the turbulence. It took a while to solve this dilemma and it didn't come directly from altruistic scientists looking for the origin of the universe. It came from practical scientists trying to hold off the end of the universe, as they knew it.

Send in the cavalry!

Back in Chapter 1, I mentioned how adaptive optics got started when the military got involved. It's now time for a little more of the story. Imagine a few thousand ballistic missiles traveling from one part of the globe to another part. Each one has a few nuclear weapons and they will strike their targets in about 20 minutes. This doesn't look like it would have a happy ending.

---

[1] Arguable, of course. Don't go there.

Now imagine if there were a way to stop those missiles. From the mid-1970s to the early 1990s, this was the mission of hundreds of American scientists and engineers.[2] Because missiles fly very fast, it is important to defend against them quickly. One cannot wait to launch airplanes or send fleets of ships around the world. Twenty minutes is a very short time. One of the ways to stop the missiles is to blow them up near their launch point before they get rolling. *Boost phase* destruction is very hard to do because we normally aren't allowed to have an array of defensive weapons deployed right next to an enemy's offensive weapons.[3]

The result of this dilemma was to build a weapon that could get near the launch sites of the bad guys really quickly and then destroy the bad guys' missiles. The only weapon that can move that fast is a beam of light. One of many schemes was to have high-energy lasers in space, a very costly endeavor, in addition to being a treaty violation. Another of the many schemes was to have high-energy lasers on the ground in the United States, and then relay the light through space with a whole bunch of orbiting relay mirrors.[4]

It was hard to build the big lasers and even harder to figure out how to get the intense beams, and I mean really intense beams, through the atmosphere. The lasers would have to have a lot of power to go as far as halfway around the Earth and then have enough power left to focus onto the side of the missile. The actual values[5] for the laser power remain classified by the military, so I looked for them in the *New York Times*.

---

[2] I, being an American scientist from the mid-1970s, do not apologize for writing this from the viewpoint of an American scientist. I would not be able to do a very good job of writing it from the viewpoint of a recluse living in a giant redwood.

[3] Oh, golly gee. That would make a nice treaty, wouldn't it? "Here's my bomb. You can blow it up if you think that I might use it for any nasty purpose that you don't like." Really …

[4] The laser light show would be something to see!

[5] Let's just say that they are *way* brighter than your average laser pointer.

To shine a laser up into space to an orbiting relay mirror meant that the light had to propagate (science jargon for "go") upward through the turbulent atmosphere. To sense the atmospheric turbulence, we need to have a wavefront beacon shining down at the exact location where the high-power laser was to hit the relay mirror. One of the crazy ideas was to have the relay mirrors at geosynchronous orbit.[6] Even though the relay seems to be stationary, it is really moving, to stay synchronized with the spinning Earth.

The speed of light is fast, but not infinitely fast. The downward wavefront beacon light takes about 100 milliseconds to come down and go through the atmosphere and our upward laser takes another 100 milliseconds to reach the relay. Meanwhile, the relay satellite, which is zipping along at almost 7,000 miles per hour, has moved. We're shooting at a moving target. Our upward laser has to *lead* the target by a half mile so that the light gets there when the relay does. We obviously can't use a wavefront beacon on the relay mirror satellite.

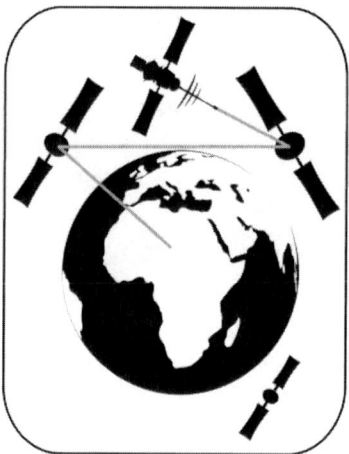

A high-energy laser is relayed through space to a target. This system has not been built yet ... as far as I know.

---

[6] An orbit 22,000 miles from the Earth above the equator. The satellite seems to hover at one point in the sky as the Earth revolves about its axis. Satellite TV transmitters are in this orbit, so that your mini-dish receiver needs to only be aimed once, rather than being tracked all over the sky.

Besides pesky atmospheric turbulence, there is another problem with high-energy lasers. The problem has a serene-sounding name: *thermal blooming.* The atmosphere, made up mostly of nitrogen, absorbs light and becomes hot. A high-energy laser going through the atmosphere (even a nonturbulent one) will heat up the air in the beam.[7] If the beam is focused at some spot in the air (presumably at a target), the air heats up and expands like a hot air balloon. The index of refraction goes down and the spot of heated air acts as a lens that defocuses the beam. The result is a focused spot that gets larger and looks like a flower blooming.

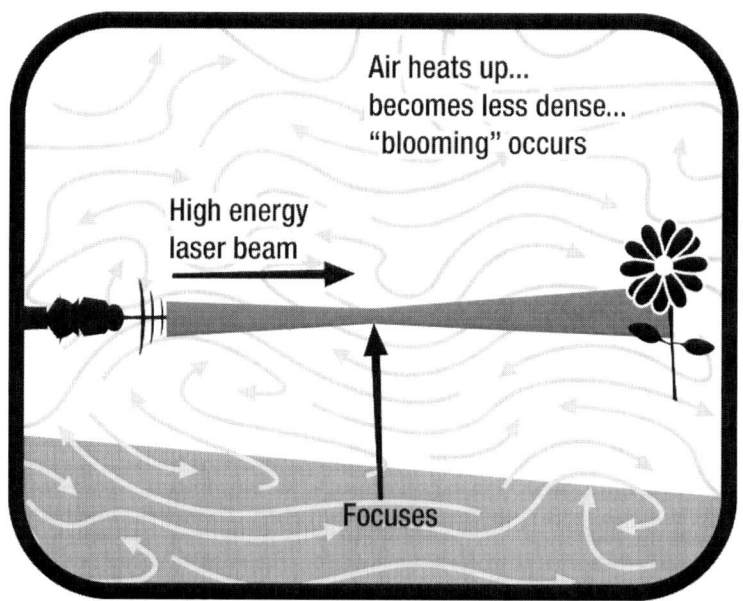

A high-energy laser beam gets absorbed by the air when it is focused. The air heats up and becomes less dense, which defocuses the beam. The laser spot seems to get larger, like a blooming flower. More focusing just makes it worse.

---

[7] It doesn't have to be a laser beam. Radars heat up the air with their beams as well. Slowly sweeping radar beams, like those used for airplane tracking, heat up the air in the beam. Birds have been seen following the radar beams just to stay in the warm air.

But, fortunately, we are no longer blooming idiots. We understand the phenomenon. Unfortunately, adaptive optics won't help. If our very smart, but not very clever, adaptive optics system sees a defocused spot, it will want to focus it some more. That means it will put more energy into the small spot, causing it to bloom even more. More blooming begets more focusing which begets more blooming.[8] We cannot fix this with conventional adaptive optics.

Fortunately, for our blowing-up-an-enemy-missile problem, thermal blooming is not going to do us much harm. The missile is moving so fast that the high-energy beam is constantly tracking it and moving along with the missile. It doesn't stay in the same bit of air long enough for absorption to be a problem. However, we still have that pesky atmospheric turbulence.

If we can't put a wavefront beacon on our own relay, and the relay is moving so fast that we can't use a fixed background star, what can we do? Many different proposed conceptual solutions started to appear in classified work around 1980. The surviving concept, suggested by Julius Feinleib, was to make an artificial wavefront beacon high in the atmosphere.[9]

Scattering is the solution

We know that the atmosphere absorbs light. It also scatters light. The scattering phenomenon is named *Rayleigh scattering* after the third Baron Rayleigh.[10] Anyway, the idea caught on—the idea of using a laser of medium power to stimulate some scattering high in the atmosphere that could be used as an artificial wavefront beacon (or laser guide star). Even though the laser guide star concept was classified, work quickly got under way to build one and demonstrate it.

---

[8] While scientists call this a *nonlinear process*, I call it a pain in the *aperture*.

[9] The concept was immediately classified by the U.S. military. Go look for the citation in the *New York Times*.

[10] Baron Rayleigh is also called Lord Rayleigh. His given name was John William Strutt. "Strutt scattering" or "struttering" doesn't quite roll off the tongue.

In principle, it sounds really great. Just shine a beam up into the air, exactly where you want your beacon to be, and presto! a wavefront beacon. However, as in all fun research, it isn't so easy. First, you will get scatter all along the path, not just at some point in the sky. [11]

A laser forms an artificial guide star in the same direction as the science object. The "darker" light from the laser beacon follows the same path through the atmosphere as the "lighter" light from the science object.

The wavefront beacon is now a line, not a point. While this is easy to fix by just applying a shutter to catch light from only one altitude (like the shutter on your camera), there are still problems with the Rayleigh guide star. The biggest problem arises because the air above us has different densities at different altitudes; there is more air near the ground, so we can get more scatter here. But if we sense the atmospheric turbulence only near the ground, we miss what's happening at high altitudes.

---

[11] Look at your headlights in the fog. Optical scattering is all over the place.

This so-called *cone effect* has nothing to do with ice cream or characters from the old *Saturday Night Live* TV program. The region that we sense is only a small cone of air above our telescope.

The cone effect. The only part of the atmosphere that the laser guide star light shines through is a cone of air between the telescope and the guide star. It misses what is above the laser guide star. The seven-year old sister of Barbara is seen in the picture.

If we create the laser guide star (Rayleigh beacon) at a high altitude, there isn't enough air to create a bright enough beacon. You see the dilemma? Too low = bright star, but missing most of the atmosphere. Too high = getting all of the atmosphere, but too dim. Fortunately, we found a happy medium. She[12] was able to use her Tarot cards to predict that we had enough laser power available to get bright enough scattering at an altitude of 20 kilometers, not at the very top of the atmosphere, but high enough to sense the highly turbulent jet stream at 10 kilometers.

---

[12] Madame Fortunatica, a rare medium who was, well, done.

Thank you, sodium

That seemed like a good solution, but it wasn't good enough. We really would like to get around that annoying cone effect entirely. Luckily, science, and a clever mind, prevailed. Will Happer, a Princeton University scientist, had been studying the very upper layers of the atmosphere, the mesosphere (a region 90 kilometers high). We're not talking about air here, but there are things going around the Earth on the edge of space.

I am going out on a limb by saying that we are lucky to have meteors hit the Earth. A mere 212 million years ago, a really big one hit and wiped out something like 95% of all living species. That would be considered lucky to a very few. Then, 65 million years ago, another really big one hit and wiped out the dinosaurs, while it left the planet to the dominance of little hairy creatures, lucky for us. But every day, thousands of little bitty meteors hit the upper part of the atmosphere and burn up before doing too much damage. Fortunately, for adaptive optics, these meteors contain all kinds of really neat chemicals, like potassium and sodium. Sodium is special. For example, it's one half of sodium chloride, table salt. Salt is a very stable compound. My doctor tells me to avoid salt, so I don't tell him that I really do like French fries.

Back to the mesosphere ... when the sodium in the meteor hits the atmosphere up there, much of it remains as just sodium, not salt. This form is called *atomic* sodium and it has a nice characteristic. The little atoms of pure sodium can absorb a beam of light if it is just the right shade of orange.[13] Then, a fraction of a second later, it emits the light again with exactly that same color. This neat little trick is called *resonant backscatter* and we can use it for a really high-altitude wavefront beacon. An artificial wavefront beacon or laser guide star that comes from 90 kilometers high acts almost as if it is coming from really far away; there is no cone effect.

---

[13] The color is represented by its wavelength, 579 nanometers (abbreviated as 579 nm).

The only remaining problem is to build a laser at the magic color 579 nanometers and focus it at the sodium up at 90 kilometers. This is not as easy as it sounds, but a lot of really clever scientists, engineers, and latter-day alchemists have done it.

It's still fun to look at the difficulties of making a sodium laser guide star.[14] First of all, there is not much sodium up there. If you take your little spaceship up and gather all the sodium, you could fit it under your desk. I wouldn't try this at your desk at home. Sodium, by itself in atomic form, is flammable[15] and explodes if it comes into contact with water.

For the laser guide star, this not-much-sodium is a serious problem. You can make a laser really powerful, but there is only so much sodium in the beam where you focus it, so your guide star has limited brightness. More laser power doesn't translate to a brighter guide star.[16] Also, the sodium layer keeps moving around. Sometimes it is at 90 kilometers, sometimes it moves up to maybe 92 kilometers. It also changes with the seasons, and sometimes moves from day to day. Plus, the number of sodium atoms in a given region varies, by as much as a factor of 10. Can you imagine having a light bulb in your house that one day is 150 watts and then a few days later, in another room, is 15 watts? I would wonder if this wasn't some new global warming legislation like The Hiding Light Bulb and Arbitrary Brightness Preservation Act.

Despite all the difficulties, sodium laser beacons are in use at a lot of the major observatories. The problem with having a wavefront beacon bright enough in the right part of the sky was solved with the development of laser guide stars. By 1991, after the fall of the Soviet Union, the classified part of the laser guide star research was unclassified. This also happened when two astronomers, Renaud

---

[14] Fun for me. I've never personally had to build one.

[15] Flammable is the same as inflammable, and is the opposite of nonflammable. (One of only three strange things that I've noticed about the English language.)

[16] The phenomenon is called *saturation*.

Foy and Françoise-Claude Labeyrie, had figured out the idea on their own and made it public.[17] It was no longer necessary, or even practical, to keep it a military secret. Area 51 was opened to the public.[18]

A sodium guide star forms above most of the atmosphere. The cone effect is gone and so is Barbara's little sister. In her place is Kenneth's little brother, Spike.

The strange case of atmospheric tilt and laser guide stars

There still remains one problem with laser guide stars. When we shine a laser up into the atmosphere, we only know the direction in which we pointed it. Large blobs of air act as big tilted mirrors or glass wedges and redirect the beam. Unfortunately, one of the

---

[17] R. Foy and F. Labeyrie, "Feasibility of adaptive telescope with laser probe," *Astronomy and Astrophysics*, **152**, 129-131 (1985).

[18] In fact, it is not. The aliens and their spaceship are still hidden there.

things we want to measure with our laser guide stars is how much the atmosphere redirects the beam. We call this simply the *tilt* of the beam.

You can see this effect for yourself. Remove all the children and the frozen turkeys from the swimming pool. Get a laser pointer for your laser guide star. Point the laser across the surface, slightly downward into the water.[19] You see the reflection from the bottom of the pool and it seems to come straight from where you pointed it. However, the gremlins of refraction actually bent the beam when it passed through the surface. The actual reflection comes not from the direction you pointed it, but from somewhere else. This is also the reason things look magnified under water.

The light bends exactly the same way at the surface of the pool, whether it is traveling into the water or out of the water. The laser pointer spot appears to be where you point it, rather than at the actual place it hits on the bottom.

---

[19] Don't point it straight down. The effect won't work. And you might drop it in the pool, ruining the laser pointer.

In any case, we don't know where the reflection came from. The amount of bending totally depends on the actual index of refraction of the water. If we don't know how much it even bends the beam, we can't measure the index of refraction, which tells us the tilt of the beam. The same thing happens when we form a laser guide star in the upper atmosphere. We don't ever really know where it formed. We just see it where we pointed the laser. Plus, the big blobs of air are moving around and changing a few times each second, so we can't measure the wavefront tilt with an artificial laser guide star.

What to do? What to do? We could choose science objects that are reasonably close to a natural star to get the tilt information and get the other more detailed information from the laser guide star. And now, I hear Barbara and Kenneth yelling at me: "Well, dummy, if we had a natural guide star already, why did we create a laser guide star in the first place!?"

We're lucky here. Physics is going to help us once and for all. The brightness of the light needed for the natural guide star to measure tilt is *much less* than what we need to measure the other things, like focus and astigmatism. We can use a fairly dim natural guide star (and there are still millions of them around) for tilt and use our laser guide stars for the other things.

Summary of the fifth chapter

Because a wavefront sensor must have a bright beacon or source of light to work, artificial laser guide stars are used. The guide star must be close to the science object[20] because of anisoplanatism. If the guide star is too low, the wavefront sensor only measures the small cone of light between the telescope and the guide star. The cone effect is avoided by placing a sodium laser guide star at a very high altitude. Beam tilt cannot be measured with a laser guide star because we don't really have control over exactly where the guide star will form.

---

[20] Nearby in the sky, not physically close to it.

In the movies, Kenneth went to get some popcorn and forgot which one of the 49 theaters was showing his movie. He fell asleep watching the classic World Wrestling Mania 6. Barbara never noticed.

Chapter 6

Mirrors That Get Bent Out of Shape

Deformable mirrors are fairy tale devices that can somehow move with magical pixie dust to change the wavefront of the beam of light that was all distorted by atmospheric turbulence whose name is the Big Bad Wolf. At least that's the scientific explanation.

For some people, especially those who are just learning about adaptive optics, the deformable mirror is just the coolest thing on the planet. Some call it a *rubber mirror* because it can bend somewhat like a rubber mirror would bend if by chance you could make a mirror out of rubber.[1] Others call it an *adaptive optic*, thinking it is the optical gizmo that is actually doing all the adapting. But, as I explained in this book, adaptive optics is more than just the deformable mirror; it is all of this stuff. And I have vowed that, from this day forward, there will be a curse placed upon anyone who seriously contemplates calling a deformable mirror an adaptive *optic*.[2]

A deformable mirror is actually quite a simple concept, much like a house cat. They both lie around and do nothing until they are aroused by some external stimulus and then they move, often quite rapidly, from one end of their immediate world to another. Then, they go back to doing nothing.

---

[1] Rubber is not very shiny, even in a Goodyear.
[2] The specifics of the curse are too graphic to write into a footnote.

Deformable mirrors, unlike cats, are also very easy to make.[3] Take a perfectly good flat mirror, just like the one in your bathroom, and deform it. I did this exact thing this very morning. I removed it from the wall, very carefully. I placed it, shiny side down, straddling the sides of my bathtub. Then, with gentle pressure across the center of it, I deformed it. It was no longer flat. I had a deformable mirror.

This happened just as the housecat that I was talking about got into one of her "active" moments where she raced around the house, seemingly without any purpose whatsoever, and jumped up on the back of the mirror where I was applying pressure. Together, we reached what is known to mechanical engineers as *fracture stress*. The largest remaining piece of my bathroom mirror is a jagged triangle about 4 inches across. It is now hard to even see my face to shave.[4] It is even harder to explain to family and friends that this was a scientific experiment.

In a real adaptive optics system, when you don't have the interference of felines, you will find a mirror that is deformed into a precise shape that will match the turbulence-deformed wavefront. That way, when the light coming from our big telescope to form the image bounces off the deformable mirror, it gets straightened out. The final image doesn't show the distortions of the turbulence. Because the atmosphere is changing all the time, the deformable mirror must be deformed differently as fast as the atmosphere changes.

What confronts us is this: We need a very good mirror when it is not deformed and then we need a way to deform it precisely at a lot of locations over the surface and do it very fast. Oh, and it has to be reliable and cheap and not break.[5]

---

[3] This is where all the engineers that have ever actually made a deformable mirror are putting a curse on me.

[4] The beard is coming along nicely.

[5] Note to self: Restrain the cat.

Segmented mirrors

The history of high-tech deformable mirrors is filled with excellent designs, a lot of innovation, and a lot of scrap. The first mirrors that were tried were really not individual deformable mirrors, but an array of small mirrors, each one capable of independent motion. The *segmented mirror* had the little segments arranged closely together to make the whole thing look like a continuous mirror. It was unfortunate that even the best ones that could be used for image-forming applications in observatories were pretty much worthless for high-energy laser beams. The gaps between the segments, even though they were thinner than a human hair,[6] allowed high-energy laser radiation to get behind the segments and vaporize just about everything. Segmented mirrors were not used for high-energy lasers after that.

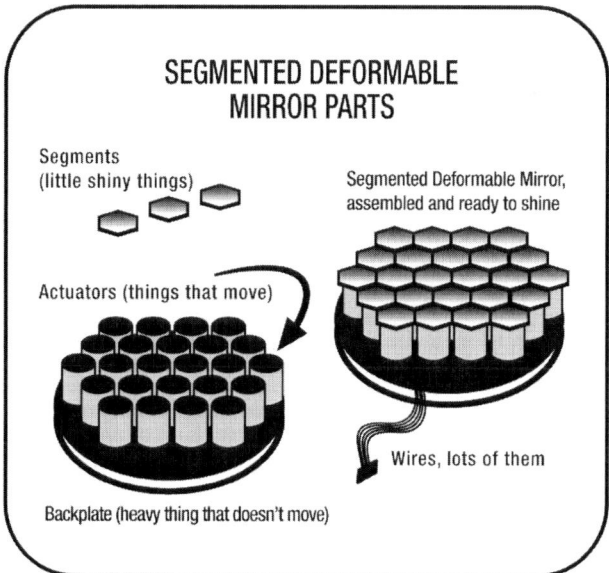

The innards of a segmented deformable mirror.

---

[6] This analogy is used because everyone knows how thin a human hair is. Some are even getting thinner as you read this. Remember the guys in the wedding photo in Chapter 2.

Continuous-faceplate deformable mirrors

The next development was a *continuous-faceplate* deformable mirror. This was fundamentally similar to my bathroom mirror, before the cat incident, except that it had plungers glued to the back of the shiny surface (the faceplate). The plungers are technically called *actuators*. Each actuator could be activated[7] by a voltage or another electrical signal, and the mirror would deform right where the actuator is glued on. The shape that the mirror surface takes around the position of the actuator is called the *influence function*.

When you have a lot of these actuators glued all over the back of the mirror surface, by applying various voltages to each one, you can deform the mirror into virtually any shape, which is what we want to do. The back of the actuator is glued to another plate that doesn't move, called the *backplate*. The wavefront sensor will tell us what shape to become and the deformable mirror will go there.

It is very important that this happens quickly. Remember that the atmosphere is changing about 20 or 30 times a second or faster. The actuators, which don't have to move very far (actually only a few wavelengths of light),[8] have to get where they are going and not get confused or slowed down on the way.

Early actuators were made of piezoelectric[9] materials. Piezoelectrics have the nifty characteristic that, when a voltage is applied, they expand. If they are glued to something that can't move (like the backplate) then the expansion causes the mirror to deform. Cool.

---

[7] *act*uator, *act*ivated, get it?
[8] About 10 millionths of a meter.
[9] Bricks inside a pizza oven.

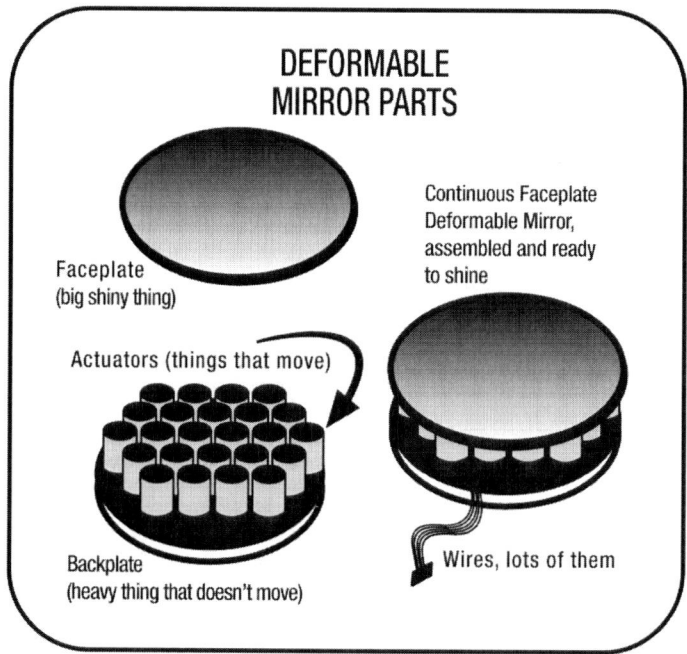

The innards of a continuous-faceplate deformable mirror.

Piezoelectric materials have some drawbacks. The first is that they are not very efficient. They need a lot of voltage (thousands of volts on each one) to move much at all. The second drawback is that they exhibit hysteresis.[10] When the hysterical actuator is commanded to go somewhere, it often goes right there, but then, when it is commanded to go back to where it started with a negative command, it doesn't quite reach the starting point. You have to give it a boost with another negative command. This slows it down. And you are never quite sure where it is at any particular time, nor will it ask for directions.[11]

---

[10] The word *hysteresis* comes from the Greek word *hystera* meaning *to be late*.

[11] Kenneth says that this is just a "guy" thing.

Vast improvements have been made on actuators since the introduction of the first very hysterical ones. Other chemical compounds with less hysteresis and more efficiency have been introduced. There is even a class of materials called magnetostrictive,[12] that don't need high voltages. Unfortunately, they do need high electric currents.

Bimorph mirrors

The idea of plungers working perpendicularly to the surface is still very popular, but there is another clever way to make a mirror deform. Let's form a mirror with one shiny material on the top and another sheet of another material glued to the back. You can try this with one piece of white bread and another piece of whole wheat bread glued into a sandwich without any meat.[13] The arrangement is called a bimorph[14] mirror. If a voltage is applied to one of the materials (the white bread), then it will expand along the same plane as the surface. If the other material (the whole wheat) doesn't expand, the whole thing will bulge up in the middle. Try it with your glue sandwich. You can use toothpicks instead of glue. Just press on the opposite sides of the bread and watch it bulge up. You can also leave the bread out for about a week and watch the mold deform the bread, but this concept, as far as I know, is not being used in any earthly adaptive optics systems.

The one really nice thing about the bimorph mirror is that the bulge takes exactly the same mathematical shape as the curvature measurement from Roddier's curvature sensor. In this case, the reconstructor (see the next chapter) can be greatly simplified.

---

[12] The expansion of the actuator is caused by a magnetic field which is caused by an electric current running in a coil around the actuator.

[13] Do not eat this. I repeat, do not even think about it.

[14] Two (bi) things forming (morph). Greek again.

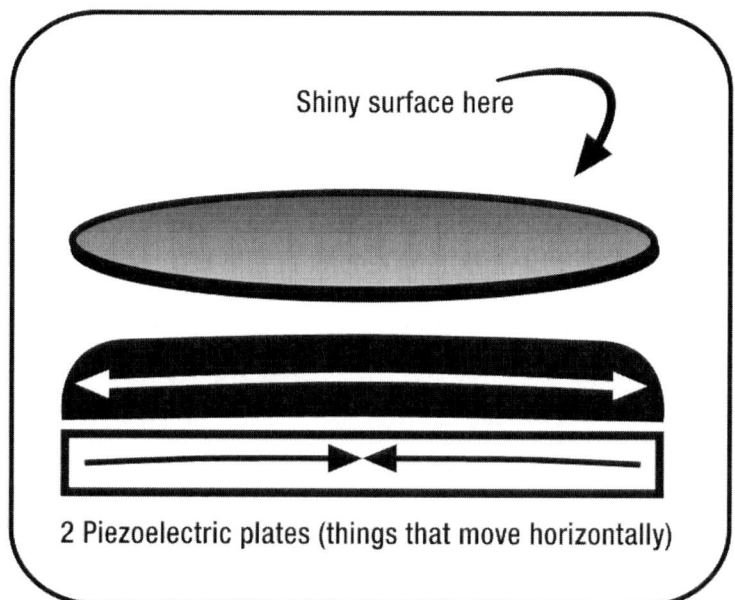

Shiny surface here

2 Piezoelectric plates (things that move horizontally)

A bimorph mirror has two plates that can move horizontally with respect to each other. (Their respect for each other has limits, though. It ends when one of them forgets to put the cap back onto the toothpaste.) Seriously though, there is one pair of plates for each independent area of the mirror, much like the way there is one actuator for each somewhat independent area of a continuous-faceplate mirror.

Micro-electro-mechanical systems

Over the past decade or so, micro-electro-mechanical systems (MEMS)[15] have been used to build all kinds of magical little things. By using the same technology that is used to build microcircuits in computers, engineers have built all manner of mechanical gizmos and electromechanical thingamabobs at a microscopic scale.[16]

---

[15] I refuse to write out "micro-electro-mechanical systems" any more.

[16] Microscopic refers to things that can be seen with the naked or clothed eye, but their innermost details are only visible under a microscope.

They've built on a microscopic scale entire working engines, using simple machines (levers, inclined planes, screws, wheels and axles, and springs) all thinner than a human hair.[17] Rumor has it that some are building entire cities without crime, pollution, or speed bumps.

One of the really cool devices is a MEMS deformable mirror. Various small companies, and spinoffs from large companies, have built segmented mirrors, continuous-faceplate mirrors, and even just tilt-only or focus-only corrector mirrors.[18] They all work fine, but each manufacturer has a number of secret designs, patents, and special features that set each mirror apart. Most of them, though, work pretty much like the large deformable mirrors, only they are smaller. I'll describe the continuous-faceplate MEMS deformable mirror (MEMS DM) and just let you guess about the rest of them.

The actuators are not piezoelectric, nor magnetostrictive, nor glued whole wheat bread. They are electrostatic. They use the same process that makes your hair stand on end when you walk across a carpet on a cold, dry day. This works even better with thin hair.[19] It seems that there are two types of electrical charges: positive and negative.[20] Opposite charges attract each other; like charges repel each other.[21] In the MEMS DM, the shiny faceplate is grounded or placed with a negative charge. Behind each region of the mirror, instead of having glued-on actuators, the DM has a little electrical pad, very close to the faceplate, but not touching it.

---

[17] Enough about the thinning hair!

[18] I am indebted to the many manufacturers of these devices. I have been able to test and perform numerous experiments with most of them. However, I won't acknowledge your names here because you keep changing them or you go on getting acquired by megacompanies while you walk away with millions. I haven't seen any of *that* yet.

[19] Oh come on, not again!

[20] I am positive that this is the case.

[21] This was not a personal discovery of mine. Also, I'll refrain from using the birds and bees analogy.

When a little positive voltage is applied to the tiny pad, the pad and the faceplate are attracted to one another, like Barbara and Kenneth were at one time. This attraction[22] causes the faceplate to deform a little bit. The result is a fully functional deformable mirror with many actuator pads jammed into a tiny area. Mirrors of this type have apertures (sizes) of a few millimeters to a few centimeters and can have hundreds of actuators across the mirror surface. Knowing full well how the price of computer chips goes down with large-volume manufacturing, the promise of low-cost deformable mirrors is on the horizon. The only current problem is that the large volumes have not been reached yet, and new innovations seem to propel the technology forward, rather than stay on a single design and just make a lot of them.

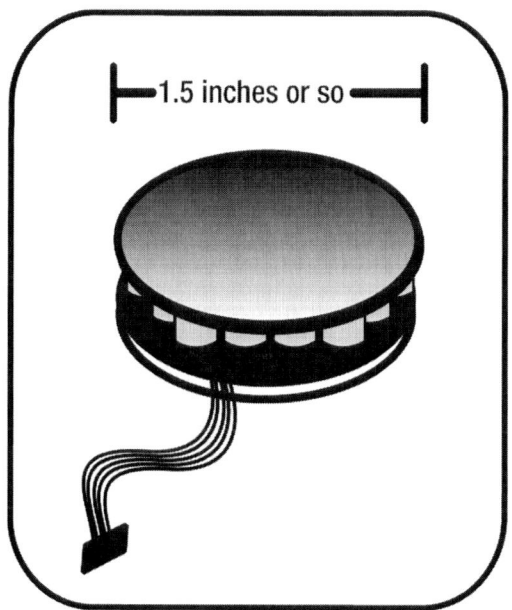

1.5 inches or so

A MEMS deformable mirror is much like the continuous-faceplate mirror or the segmented mirror, except that it is really, really tiny.

---

[22] On the deformable mirror, not between Barbara and Kenneth.

All of this can be attributed to the unfortunate fact that adaptive optics is still not a viable solution to curing disease or hunger in the developing world. But, it's important to end the discussion of deformable mirrors on a high note. I choose the "B-flat" above "high-C," the highest note that I can sing without requiring surgery.

Summary of the sixth chapter

Deformable mirrors are fairy tale devices that can somehow move with magical pixie dust to change the wavefront of the beam of light that was all distorted by atmospheric turbulence whose name is the Big Bad Wolf. At least that's the scientific explanation.[23] Deformable mirrors come in a few different types, such as segmented, continuous-faceplate, bimorph, or little bitty MEMS mirrors.

Barbara and Kenneth don't remember much of this chapter either.

---

[23] This isn't so much of a summary as it is a repeat of the first paragraph. If you didn't read the rest of the chapter, you missed the part about the cat.

Chapter 7

Computers That Shouldn't Crash

Wavefront reconstructors are one of the three major subsystems of an adaptive optics system. Wavefront *reconstructors* sound like they should be massive machines that roll up and down the beach to make sure surfers and bathers have the time of their life. What wavefront reconstructors actually are is not as exciting as the scene just described.

They are simply computer-like things that take the measurements of the wavefront sensor and get it ready to create a correction for the distorted beam of light with a deformable mirror. In some cases, they are pretty complicated computers, working blazingly fast. In other cases, they are simply wires that hook the wavefront sensor to the deformable mirror.

Even though it's not the order in which the adaptive optics subsystems go together, it's never easy to discuss the reconstructors before a lesson about how deformable mirrors work. If I reversed the chapters and discussed reconstructors first, then you would have Chapter 7 in front of Chapter 6 and it would make sense, but really confuse the copy editors. I choose not to confuse the copy editors; they always have the last word.

The job of the wavefront reconstructor is to decode the stuff from the wavefront sensor. What the wavefront sensor sends to the reconstructor is very important. With a Shack-Hartmann sensor, it would be the position of the little Hartmann spots, maybe a different voltage for the position of each spot. It is usually two voltages for each spot, because the spots can be left or right *and* up or down. Or the wavefront sensor might send a digital signal (bits, bytes, that sort of thing that computers can read) that represents the position of the spots.

If the wavefront sensor is a shearing interferometer or pyramid sensor, the same thing is possible; voltages (analog) or bits and bytes (digital) represent the local tilt of the wavefront all across the optical pupil. If the sensor is a curvature sensor, the voltages may represent the amount of wavefront curvature at each position. In any case, there may be an awful lot of numbers (or voltages) to contend with.[1]

A daunting problem

Let's put this all into context with an example. Say that Prof. Smythe has a 10-meter telescope.[2] Our wavefront sensor has 2000 subapertures (the little lenslets) that are measuring the little wavefront tilts (4000 of them[3]) at about 300 times every second. Let's also say he has a deformable mirror with 2000 actuators, each one requiring a single voltage to tell it where to go. So, as fast as he can, he must calculate the 2000 actuator voltages from the 4000 wavefront tilts. There is never an easy solution to this.

This is typically done by using a method called *matrix multiplication*. The calculation is possible because we will assume that our adaptive optics system is *linear*. All that gobbledygook means that when we

---

[1] "...with which to contend." is better grammar, but who talks that way?
[2] We would all like one. I asked Santa for one last year. Instead, I got a piece of coal.
[3] Left-Right,Up-Down for each one.

measure something with the wavefront sensor or send a voltage to the deformable mirror, things act proportionally.

Say that the wavefront has a little tilt at a particular subaperture; its value is 4.[4] That will cause the output of the wavefront sensor to be a specific value, say 6. These are arbitrary numbers right now. Instead of 4 and 6, it could be 548 and 2141, but don't get upset. We say it is *linear* when these numbers are proportional to one another. For example, if the little tilt (previously 4) is doubled (to 8), then the output of the sensor (previously 6) will also be doubled (to 12). Aw, just forget it. It works really great if it's linear and it gets really complicated if it is nonlinear.

The same thing happens with the deformable mirror. The voltage to each actuator is proportional to the amount of motion that changes the surface. The voltages are sometimes called *commands* because that is the way we command the mirror to do something. We do not suggest it; we do not prod it or coax it; we don't threaten it with a time-out; we do not even act very nice to it; we command it!

Because of this linearity assumption,[5] we can just multiply each of the wavefront sensor signals (one for each little tilt measurement) by a bunch of numbers to get one of the deformable mirror commands. The bunch of numbers, mathematically speaking, is arranged in a special way called a *matrix*. Remember, the output of the wavefront sensor gives us voltages or bits and bytes that represent each little wavefront tilt. In a computer somewhere, usually close by and not in the next county, we multiply the numbers by a matrix of other numbers to come up with the deformable mirror commands. The matrix then reconstructs the wavefront to form the correct deformable mirror surface. The matrix gets its own special name: the *reconstructor*.

---

[4] 4 what? Degrees, microns, furlongs, it doesn't matter. Just stay with me.
[5] The linearity assumption is just a very good first guess. Don't hold me to it forever.

Marriage counseling may work

Remember Barbara and Kenneth, who got married in Chapter 2? Things are not going so well. They had a spat, a row, and a quarrel all rolled into one big ol' shoutin' match. Barbara was discussing the future of their relationship while Kenneth was wondering what the Redskins were going to do with a fourth-and-24. By the time the game was over, Kenneth was opening his last beer (before lunch) and Barbara was back living with her parents. After a few weeks of agonizing, Barbara contacted Dr. Mannheart, the compassionate marriage counselor who got a warrant for Kenneth to appear in a counseling session. Dr. Mannheart spoke with both Barbara and Kenneth to diagnose the core of the division within the relationship. Barbara said that she was just a little upset that Kenneth never did any chores around the house and would always leave the toilet seat up. Kenneth responded that he didn't understand why on Earth the Redskins would throw a screen pass on fourth-and-24.

Dr. Mannheart sensed the problem and did a few calculations on his pad of paper. He asked Barbara and Kenneth to agree on a few simple actions to reconstruct the marriage. Barbara was commanded to take a little more interest in football and Kenneth was commanded to learn how to open a carton of milk.

Isn't this the best analogy for an adaptive optics system that you've ever read? Maybe not, but Dr. Mannheart is the wavefront sensor that diagnoses the problem. He then figures out the solution, like a reconstructor. And finally, Barbara and Kenneth are commanded to change, like the deformable mirror, which will fix things. It is still important that the deformable mirror responds precisely to its commands in order for everything to work. In the case of Barbara and Kenneth, this didn't quite happen. Barbara did take an interest in football and is now dating the Redskins quarterback. Kenneth has become an obsessive-compulsive who sits on the bathroom floor all day, and moves the toilet seat up and down, up and down, up and down.

Summary of the seventh chapter

Wavefront reconstructors connect the output of the wavefront sensor to the input of the deformable mirror. All the pieces have to work together to make the adaptive optics system work.

Barbara and Kenneth had a fight and Dr. Mannheart suggested a remedy. Barbara ran off with the quarterback and Kenneth had a nervous breakdown.[6]

---

[6] I actually find this amusing. It contributes to the end of the story, which, only I know at this point.

Chapter 8

Other Ways to Do It

Image sharpening

Now, disregard everything I've said so far about the adaptive optics system having to have a wavefront sensor, reconstructor, and deformable mirror. I lied. There are some very clever systems that don't really have a wavefront sensor. They never measure the wavefront. They don't even try. There is one kind of system that just looks directly at the image and fixes it. The concept is called *image sharpening*.

Because you have read this far, you know what binoculars are. As you look through the binoculars[1] you probably see something that looks like what you want to see, but it is blurry. It's out of focus. Somewhere on the binoculars is a focus gizmo. It may be a little wheel that you twirl with your fingers or it may be a lens that you can pull out or push in. Some binoculars will have a motorized focus contraption powered by a nuclear reactor, but these are rare. In any case, you try to take the blur out of the image in your eye by adjusting some gadget. When the image looks sharp, you stop fiddling with the contraption and stare in amazement at the object of your attention.

How did you know the image was in focus? Answer: When it looked *sharp* to you, according to your evolutionarily developed eyesight.

---

[1] No more *pair of* binoculars, just binoculars.

A creepy guy with binoculars.

You just became a *closed-loop* adaptive optics system. Your eyes (acting in conjunction with a reasonably functional brain) measured the wavefront by looking at its effect on the image. You never really measured the wavefront, but you looked at its effect. Your nimble fingers adjusted the focus just like the actuators on a deformable mirror do. A moving lens is really the simplest form of a deformable optical element. It doesn't deform itself; it just changes the point of focus. If the image got more blurry, your fingers went the other way to fix things. This was the *adaptive* part. You had a closed-loop system because you kept going until you got it right, constantly looking at the effects of your finger actions.

A computer can do the same thing as your eye-brain sharpness "sensor." Mathematical measures of sharpness range from "integral of the square of the intensity over the image" to "well, it's not so crummy anymore."

When you are finished looking through the binoculars, take them out on a sunny day, point the big lens end toward the Sun, and fry up some ants.[2]

---

[2] Hey PETA, I'm just joking.

Phase diversity

Phase diversity is another non-wavefront-sensor adaptive optics technique. It gets its name *diversity* from the social rules on how the system is assembled.[3] Rule 11: For every three *convex* lenses, there has to be at least one *concave* lens; Rule 94: If you are considering using a prism, you also have to think about using a grating; Rule 131: You must provide equal or equivalent polishing techniques to *all* mirrors, regardless of how they function; and Rule 260: You must achieve phase diversity regardless of the consequences.

Actually, the term *phase diversity* comes from the fact that there are two (or more) images recorded every time you want to apply a correction with a deformable mirror. One is a focused image, or the best that you can get so far. The second image is of the same thing, with the same optical system, except for the addition of a precisely known phase change, such as a known defocus. The second image is called the *diversity image*. By magically subtracting the focused image from the defocused one in a computer, and trying all kinds of different solutions, the best image can be created with a side effect that shows what the wavefront actually is.[4]

Multidither and other hill-climbing doodads

Multidither techniques are not used for fixing up images, but they are very useful in trying to get a good laser beam to be better. For example, every laser is somewhat screwed up. None that I know of are perfect.[5] Some applications, like laser communications, with laser beams sent through incredibly thin optical fiber, need to have a way of getting the most light into the end of the fiber and not

---

[3] The following is political satire. There are no social rules for optical systems.

[4] Other side effects include bloating, nausea, fungal growths, and plagues of locusts.

[5] You may disagree with me, but if you have a perfect laser, please introduce it to me at the next meeting of the Regional Procrastinators Society that will meet ... soon.

wasting it. Adaptive optics can help here and you don't need a wavefront sensor. Just measure the amount of light going into the fiber[6] and adjust the wavefront with the deformable mirror to maximize the light going in.

Without a wavefront sensor, how does the reconstructor know what to tell the deformable mirror to do? That's actually easy. Just try all the possibilities until it gets to the maximum. The only hard part is deciding when to stop. A lot of techniques are used to do this. Each of them takes a little time, so you want the fastest technique. If something is changing the laser, like vibration or even mild turbulence, while you are trying to fix it, you can't find the right solution.

Various things can be tried. Remember, what you are trying to do is maximize the amount of light getting into the fiber. The measure of this amount is called the *figure-of-merit* (FOM).[7] Try all possibilities and see which one gives you the best FOM. For instance, move each of the actuators just a wee bit and see what happens. Then, try another actuator or another wee bit of a move and see what happens. While robust, this is never done because literally, *all of the possibilities* would be an infinite number.

Clever people over the years have figured out ways to make good guesses and get to the maximum FOM quickly. One way is called multidither because you dither (wiggle in a controlled fashion) each of the actuators all at the same time. Each one is dithered at a different rate. At the end of the fiber, where you are measuring the light intensity, you will see the light intensity varying at all those different rates. When one rate makes the intensity go up, you know which actuator that is and keep it going in the direction that made it go up. When one rate makes the intensity go down, you know which actuator that is, and make it go in the other direction.

---

[6] One way to do this is by seeing how much comes out the other end.

[7] Another acronym. I will deposit it in the Bad Acronyms Registry Repository of Oswego, Maryland (BARRoOM).

As some of us write in technical publications, when we just want to look smarter than the reader, we say we are writing *for completeness.* I will now be complete, almost, and mention other techniques. In addition to multidither, there is stochastic gradient descent; there is Nelder-Mead simplex; there is simulated annealing; and there are genetic algorithms. All of these techniques are way too complicated for me to satirize while maintaining technical perfection.

Summary of the eighth chapter

Some systems don't have actual wavefront sensors and reconstructors, but they rely on other methods to determine how the image is screwed up and how they might unscrew it. Image sharpening, phase diversity, and other hill-climbing methods are examples of these systems.

Barbara is finding out that the quarterback has his eye on one of the Redskins' cheerleaders. Kenneth is, or appears to be, heavily medicated.

Chapter 9

Putting the System Together

Now wouldn't it be nice if we could go to a local home improvement store and buy an adaptive optics kit with all the parts? We could go home, or to our observatory, and begin to put it together. Our directions would read something like this:

Step 1. Read directions completely before beginning.

Step 2a. Keep in mind that manufacturer accepts no responsibility for absolutely anything that will go wrong.

Step 2b. Our lawyers are better than your lawyers.

Step 3. Identify all parts.

Step 4. Place Tab A in Slot H and cover with one (1) drop of glue from Envelope K.

Step 5. Employee must wash hand before returning to wok.[1]

Step 6. Assembly is complete. Unscrew turbulence.

Note to Ed[2]: It doesn't always go like this.

---

[1] Actual sign seen in rest room of Chinese restaurant with, presumably, ambidextrous staff.

[2] "Ed" is not the editor. It is Ed, the guy in the next cubicle who plays Wii[TM] all day long.

Professor Smythe builds a system

Actually the design, assembly, and operation of an adaptive optics system can be quite complicated. Let's find out what Professor Smythe is doing. He must first figure precisely what it is that he wants to do with his telescope and adaptive optics.[3] Let's just say, for the sake of argument, that Prof. Smythe wants to have an imaging system for a 4-meter telescope. And also for the sake of argument, this telescope is high on a mountain where the seeing is pretty good. And finally, for the sake of argument, let's just say that the atmosphere's coherence length (size of the air blobs) is about 20 centimeters. To settle the argument,[4] Prof. Smythe calculates that there are about 320 of those 20-centimeter blobs over his 4-meter telescope aperture. Ta daa! Drumroll, please.

Prof. Smythe decides to build an adaptive optics system with 341 actuators on his deformable mirror and a Shack-Hartmann wavefront sensor with 180 subapertures.[5] A matrix-multiply reconstructor will be put into a digital computer. A sodium laser guide star is added because it would really be cool to have one. A really expensive science camera is installed to collect all the images of the births of stars, especially the ones featured at the Emmy Awards.

Prof. Smythe chose not to use a segmented deformable mirror or a bimorph. He chose not to use a curvature or pyramid wavefront sensor. He also chose the color mauve for his desk chair.

I have managed to get three-quarters of the way into this book without showing you what an adaptive optics system would look like. Here's one:

---

[3] Using it as a paperweight is not recommended.

[4] At this juncture, the designers have stopped arguing and are now involved in a game of rock-paper-scissors to decide how to proceed.

[5] There are two measurements in each subaperture and, as long as the number of measurements is more than the number of actuators, it should work, sometimes.

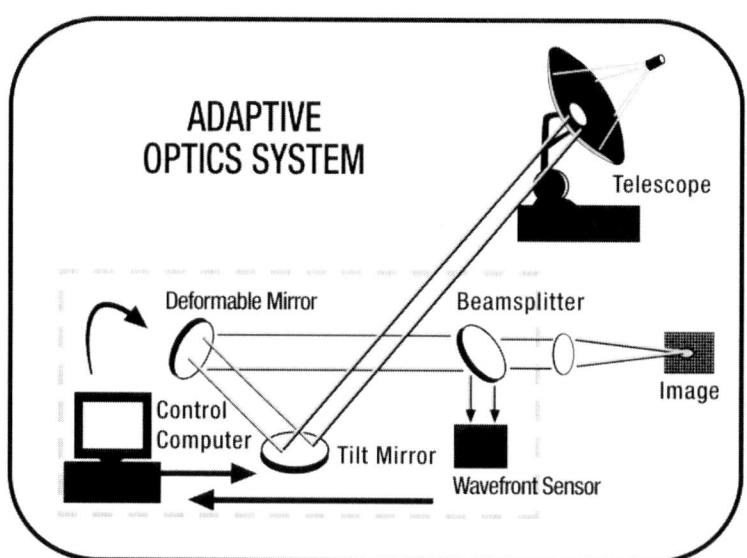

The main parts of an adaptive optics system. The mirrors for the conjugate pupils are not shown because they were too hard to draw.

The basic configuration, following the light from the star, should go something like this: the telescope primary mirror[6] (which will be a conjugate pupil of the deformable mirror); the telescope secondary mirror;[7] a tip/tilt mirror;[8] a bunch of mirrors to create a conjugate pupil where the deformable mirror needs to go;[9] the deformable mirror;[10] and a beamsplitter[11] to send a sample of light to the wavefront sensor.[12]

---

[6] The big one that first gets hit with the light.

[7] The little one that sends the beam to the correct focus at the camera.

[8] To correct for the big wobbles in the image, due to the big wobbles in the atmosphere.

[9] This is so that the DM corrects the errors in the right place. Don't concern yourself with the details.

[10] Where the wavefront correction happens.

[11] It splits the beam.

[12] If you feel completely lost at this point, think how I feel. I've been lost since that part about statistics in Chapter 2.

Now it gets messy. We have two beams of light. One goes along its merry way and gets focused (with another bunch of mirrors) onto the science camera to get the really pretty images for *Popular Science*. The other beam has to go to the wavefront sensor. However, before the light gets to the sensor, there is another bunch of mirrors that create another boring conjugate pupil at the wavefront sensor. This way, we have three conjugate pupils, each an exact optical replica of the other ones. By doing this, we measure the wavefront at the wavefront sensor, which can then be deformed into the deformable mirror, which is optically tied to the primary mirror, where we really want to correct for the atmospheric turbulence. Those are some clever conjugates.

There are a lot of other details to work out. The first one deals with geometry. How do we arrange 180 subapertures and the 341 actuators? Prof. Smythe calls this dilemma *registration*. Do we make sure that the actuators are centered at the corners of the apertures, or don't we? Do we make sure an actuator is centered on a subaperture, or don't we? In spite of the many detailed mathematically charged papers that have been written on this very topic, the answer is: It pretty much doesn't matter. As long as whatever arrangement you decide on works, it should work fine. But you must make sure that it doesn't change. The reconstructor computer is like a housecat. It lives by routine. It hates variation. Don't ever take your computer to a veterinarian.

Barbara, Kenneth, and bananas

Back to Barbara and Kenneth, who are still having difficulties. Barbara has dumped the quarterback and now is more interested in saving something. It doesn't matter what. One week it was whales. Another week it was the lost children of Camden, New Jersey. Later in that same week it was the lost children of Wales. She is searching for the meaning of life and she is thinking about getting back together with Kenneth.

On the home front, Kenneth is doing better. He has finally decided to leave the toilet seat in the down position. He recently read the cover of the supermarket tabloid *World Wacko News from Hollywood* that said that there are more bacteria on the kitchen counter than on the toilet seat. Kenneth is now cutting broccoli and bananas on his toilet seat and thinking about getting back together with Barbara.

The f-number dilemma

Prof. Smythe has problems other than bananas to worry about. To put an adaptive optics system into a telescope, Prof. Smythe has to worry about a parameter called the *f-number*. It is the f-number (the "focal number") that describes how "fast" or how "slow" the image is formed. This may sound like we're talking about time and you're right, we're talking about time. The f-number is a ratio between the focal length of a lens and the size of the lens.[13] In a system of lenses, we use all of them in the calculation. For more than one mirror or lens, it gets messy.[14]

When the ratio or f-number is high, either with a large focal length or a small lens size, (numbers like 5 or 10 or 20 come to mind), the focus is a long way from the lens. It takes a longer time[15] for the image to come into focus. We call this a *slow* system. With a small f-number, when the focus happens only a short distance away, we call it a *fast* system. Fast f-numbers are values like 0.50 or 1.25 or 2.0. So what has this got to do with adaptive optics and Barbara and Kenneth?

Barbara and Kenneth have got to "get on the same page," "see eye to eye," "find a level playing field," or another ridiculous metaphor. The point is, they have to agree on *something* before they can

---

[13] You can substitute *mirror* for *lens* here. Just don't tell Mom who did it.

[14] Consult the classic text *Getting Messy with f-Numbers* by Leslii Smythov, Potato Press, Leningrad, USSR (1931).

[15] Maybe nanoseconds, but not much more.

begin to agree on anything else. For Barbara and Kenneth, we find out that they both agree that Cheerios can reduce bad cholesterol if taken as directed. They can now go back to Dr. Mannheart to fix things up.

Our adaptive optics system has the same dilemma. It just so happens that there are two types of optical subsystems in our telescope adaptive optics system. The optical variety[16] of one of them is *verging*. The optical variety of the other is *collimated*. When a light beam is going straight and not spreading out, it is called collimated.[17] When the light beam is getting smaller it is con*verging*. When it is getting larger it is di*verging*.

The primary mirror takes (almost) collimated light from a far-away star and begins to focus it toward the image. The secondary mirror does a little more focusing and generally, the light leaving the secondary mirror is converging.

Now here comes the problem. The deformable mirror is coming up next, right after the bunch of optics that make the conjugate pupil doohickeys. The deformable mirror is usually flat.[18] It likes to work with collimated beams. To make matters even more perplexing, the wavefront sensor probably expects an input beam that is collimated.[19] And then, after the beam gets sent on its way by the beam-splitting beamsplitter, it has to again converge toward the focal plane where the image is formed.

A few pages ago when I was describing what mirrors went where in the adaptive optics system, I used the highly descriptive technical

---

[16] Optics normally do not have genders, so *variety* is chosen.

[17] Going along as in a column.

[18] It is not always flat. See the article entitled "Deformable mirrors are not always flat," by L. B. Smythe, in *Journal of Flat Things*, Vol. 3.1415926..., Oct. 1987.

[19] It does not always have to be collimated. See the article entitled "Verging wavefront sensors of the third dimension," by L. B. Smythe, in *Journal of Verging Things*, Vol. Zero, Jan. 2000.

phrase "bunch of mirrors to create a conjugate pupil." It turns out that the first bunch of mirrors must do more than *create a conjugate pupil*. They have to create a conjugate pupil and turn the converging beam into a collimated beam for the deformable mirror.[20] They have to make sure the beam has the correct f-number. The second bunch of mirrors takes the collimated beam from the beamsplitter and converges it toward the science camera. The mirrors again have to make sure that the f-number of the beam matches the f-number that the science camera expects. The sample beam going to the wavefront sensor starts out already collimated, so the only thing that its bunch of mirrors must do is the conjugate thing.[21]

The irreducible problem of multiconjugate adaptive optics

About 40 pages ago, I described the problem that we have with anisoplanatism. (Looking one direction through the atmosphere will be different from looking in another direction.) This is such a nuisance; because of anisoplanatism, the conjugate pupil can be created for only a very small portion of the sky. Adaptive optics systems with only one deformable mirror and one wavefront sensor, no matter how well built, can clearly see only a small region of the sky. The region that is clear is called the optical *field of view*. In a turbulent atmosphere, the field of view is little more than what is right around just one of those visible stars.

Sometimes astronomers want to see bigger things or a larger arrangement of smaller things. Galaxies, nebula, binary stars, and lots of other objects require a larger field of view. To get a larger field of view using adaptive optics requires many wavefront sensors and multiple deformable mirrors. Each one, along with its own laser guide star, has its own conjugate pupil. The systems get really complicated, but the multiconjugate adaptive optics system is absolutely necessary for a wide field of view.

---

[20] Age rating: 8 and up. No gluing needed.

[21] Age rating: 17 and up. May include crude language and violence.

Now get this: The number of separate combinations of laser guide stars, wavefront sensors, and deformable mirrors depends on the size of the telescope, the desired field of view, and the resolution or detail that the astronomer wants. It also depends on the budget. Things get very expensive very fast. Today (and yesterday and tomorrow, presumably) a few telescopes with multiconjugate capability have six laser guide stars and adaptive optics systems. I have heard discussions that some future telescopes may require up to 32 laser guide stars. While this sounds outrageous, we must remember that it has been only 35 years since the introduction of the computer game Pong™ and look how far we've come with computer games. We now have Grand Theft Auto IV™!

The alphabet soup of adaptive optics acronyms

It has been said that the first auto race happened as soon as the second car was built. And so it is with adaptive optics. When we had just one wavefront sensor, one guide star, and one deformable mirror, we simply said that we had "adaptive optics." With multiconjugate adaptive optics, things get more difficult to describe in a concise manner. Multiconjugate adaptive optics is abbreviated with its acronym, MCAO.[22]

This means that the old thing that was simply called *adaptive optics* is now *single conjugate adaptive optics*, SCAO. SCAO has one reference or beacon star, one wavefront sensor pointed along the optical axis (the direction that the telescope center is looking), and one deformable mirror with its conjugate being the worst part of the atmosphere, the warm ground layer.

MCAO now refers specifically to systems with multiple reference stars, some laser guide stars, and some natural guide stars, along with at least two wavefront sensors and two deformable mirrors, each one conjugated to either the ground layer or a high-altitude layer, near the jet stream at 30,000 feet (10 kilometers).

---

[22] There are apparently no rules regarding whether compound words like *multiconjugate* get one or two letters for their acronym.

Letting our imaginations run wild, we can conceive of much more complicated systems. Let's build a system with multiple reference stars spread out over the entire field of view of the telescope. We add multiple wavefront sensors looking at the light from the reference stars. The clever wavefront reconstructor calculates the best wavefront estimate for the ground layer and the one deformable mirror corrects that. *Ground-layer adaptive optics* (GLAO) is much like SCAO, except that it has multiple reference stars and multiple wavefront sensors.

This is now getting very dicey and way too confusing. However, to be complete, I'll also include a few more examples of these attempts to improve our view of the universe. Modifying GLAO slightly by putting the multiple laser guide stars closer together, our really clever reconstructor builds up an estimate of the turbulence in three dimensions—across the telescope area and all along the vertical path to the science object. Being analogous with tomography,[23] where a three-dimensional image is produced, this system is called *laser tomography adaptive optics*, LTAO. The correction is still performed with a single ground-layer conjugate deformable mirror.

If you thought that this couldn't get worse, don't worry. It does. Because SCAO works so well with a small science object with a nearby reference star, it was decided by a team of very desperate engineers that we could combine many of these SCAO systems into one telescope. Each minisystem uses the entire telescope but looks only at a narrow region (field) around each science object. There is a narrow-field deformable mirror for each object. An array of laser guide stars and natural reference stars are close enough to the science objects that anisoplanatism is not a problem. This

---

[23] Tomography is the "T" in CT scans where a three-dimensional image of our innards is made for the physicians so that they can remove the right part from the right place.

system should provide detailed images of multiple objects,[24] even those far from the optical axis where correction is always best. Of course, it has the name *multi-object adaptive optics* or MOAO.

The classic 1951 motion picture "The Day the Earth Stood Still" was scientific proof that aliens[25] from far away would come here.[26] They would convince us, using logic, reason, and the mere threat of total painful annihilation, that we should all love one another.[27] The lesson to me, however, is that I would really like to see where they are coming from—before they get here.

If we build some really big telescopes, maybe as big as a football stadium, we could see their cute little planets. That is the plan for some extremely large telescopes with clever names like the Extremely Large Telescope (ELT), the Thirty-Meter Telescope (TMT), and the Overwhelmingly Large Telescope (OWL). To get the most out of these huge eyeballs, we need *extreme adaptive optics* (XAO or ExAO—nobody has won the acronym war yet). The systems, as envisioned by those who have such visions, will be much like a sophisticated SCAO with 10 times as many actuators as we have now and speeds up to six times higher than we have now.

Every one of these *acronym adaptive optics* (AAO) systems is being considered or in design as I write this in 2008. And I am quite sure that there are other concepts out there. *Imaginary adaptive optics* (IAO) is where it all gets started.

---

[24] If it works as advertised.

[25] All the aliens apparently look like the actor Michael Rennie.

[26] Landing in Washington, D.C. and not being frightened away by the mysterious spectacle in the House of Representatives.

[27] The robot Gort was told "Klaatu barada nikto" which means "Plug in the adaptive optics system before you try to use it."

Summary of the ninth chapter

Prof. Smythe is putting together his adaptive optics system. He has decided on a wavefront sensor and a deformable mirror and has designed the reconstructor. He draws an optical layout of his system and builds all the components he needs. He has to worry about pupil conjugates and f-numbers and collimation and all those sorts of things, but his system is ready. He learns that there are many more complicated ways to build an adaptive optics system, such as multiconjugate systems and laser tomography systems, but he's saving for retirement.

Barbara and Kenneth are reconciling.

Chapter 10

Getting the Blasted Thing to Work Right or Even Work at All

So now we have it. We get all our mirrors, lenses, beamsplitters, cameras, and assorted mounts and put them all together. We screw them down. We have to align them. We have to register the DM and the wavefront sensor. We have to turn on our really cool sodium laser to make a laser guide star way up in the sky and then we flip the switch and … BAM!

Barbara and Kenneth enter the picture. As they fix their marital difficulties, they have found that when Barbara makes a small change in her behavior, Kenneth responds with a small change in his behavior. (Kenneth usually has to work very hard to forget about the quarterback calamity.) Looking at the outcome of all their actions, each one tries to make sure that the entire *system*, described by Dr. Mannheart as their *marriage*, is improving. Barbara and Kenneth adapt to each other and have finally agreed that neither one is right *all* the time.[1]

---

[1] Barbara is not so sure about this.

Barbara and Kenneth learn to work it out.

I don't know if this is a scientific rule, principle, theory, or law, but *nothing ever works the first time it is tried.*[2] We have to check out everything. We trace all our steps and make sure that we haven't forgotten anything. Prof. Smythe has the optics. He now has to add the adaptive part!

It's not over until it's over

Prof. Leslie B. Smythe, in his enthusiasm for merely wanting to see how the universe was formed, forgot to build the wavefront reconstructor and put it all into the computer. There is a big computational problem. The Shack-Hartmann wavefront sensor spots are generated literally at the speed of light. The spot positions have to be sensed electronically. There is a little time delay in doing this. Then the 341 actuator commands have to be calculated by the computer with the wavefront reconstructor. There is a big time delay in doing this.

---

[2] Think of the story of Noah and the Great Flood. The first "do over."

Think about the problem: we have 360 Hartmann spot positions (horizontal and vertical for each of 180 spots). Each one can be affected by the deformation from each actuator.[3] What this means is that you have to multiply 360 numbers by 341 numbers and then do some other adding and wiggling of the numbers. That's nearly 125,000 multiplications. And, we have to do this in less than 1/100 of a second.

```
98 32 54 3 65 4 32 89 4389 75 49 83 12 9 54 32 98 24 35 98 23 98 2 97 8
35 67 35 66 78 98 35 09 85 4 30 9 34 90 75 43 90 5 30 95 34 90 8 45 18 75
49 83 12 9 54 32 98 24 35 98 23 98 2 97 8 49 83 12 9 54 32 98 24 2 97 8
90 75 43 90 5 30 95 34 90 8 45 18 75 49 83 12 9 54 32 98 24 35 98 23 98
35 67 35 66 78 98 35 09 85 4 30 9 34 90 89 4389 75 49 83 12 9 54 32 98 24
35 98 23 98 2 97 8 35 67 35 66 78 98 35 09 85 4 30 9 34 90 75 43 90 9 54
98 32 54 3 65 4 32 89 4389 75 49 83 12 9 54 32 98 24 35 98 23 98 2 97 8
35 67 35 66 78 98 35 09 85 4 30 9 34 90 75 43 90 5 30 95 34 90 8 45 18 75
90 75 43 90 5 30 95 34 90 8 45 18 75 49 83 12 9 54 32 98 24 35 98 23 98
98 32 54 3 65 4 32 89 4389 75 49 83 12 9 54 32 98 24 35 98 23 98 2 97 8
49 83 12 9 54 32 98 24 35 98 23 98 2 97 8 49 83 12 9 54 32 98 24 2 97 8
90 75 43 90 5 30 95 34 90 8 45 18 75 49 83 12 9 54 32 98 24 35 98 23 98
49 83 12 9 54 32 98 24 35 98 23 98 2 35 67 35 66 78 98 35 09 85 4 30 9 34
97 8 49 83 12 9 54 32 98 24 2 97 8 90 75 43 90 5 30 95 34 90 8 45 18 75
90 75 43 90 5 30 95 34 90 8 45 18 75 49 83 12 9 54 32 98 24 35 98 23 98
98 32 54 3 65 4 32 89 4389 75 49 83 12 9 54 32 98 24 35 98 23 98 2 97 8
49 83 12 9 54 32 98 24 35 98 23 98 2 97 8 49 83 12 9 54 32 98 24 2 97 8
90 75 43 90 5 30 95 34 90 8 45 18 75 49 83 12 9 54 32 98 24 35 98 23 98
```

An artist's depiction of lots of numbers.

Even for a computer, this is pushing it. It's also very boring to talk about this in too much detail. However …

Even when we assume that this daunting calculation can be done, we still have to send the actuator commands through amplifiers to the actual actuators. This is another small time delay. All in all, we have one really hard job to do.

---

[3] Don't panic here. I'm going to tell you the trick.

A lot of people have put a lot of effort into making this whole thing achievable in the short time that we have. Faster computers, quicker Hartmann spot detectors (with less and less light), and various mathematical trickeries have all been used. One of these tricks comes from just stepping back and looking at the arrangement of the actuators and the subapertures (the registration). It turns out that many of the subapertures don't ever see any effect of most of the actuators. What is so nice about this? Many of those millions of multiplications each second are multiplications by zero! And I love to multiply by zero! Zero times anything is just zero. Wow. So, by this one little trick,[4] the whole process can be sped up and we can get the adaptive optics system to be adaptive. That is, it can be responding to a changing atmosphere and still be able to see what effect our deformable mirror has on the wavefront and the image.

The last little bit of knowledge that I want to impart to the world via this book is a method to calculate which numbers for multiplying go into the reconstructor. This is like the magic formula for Barbara and Kenneth to get together again. The numbers are aligned in rows and columns in the matrix. There are two rows for each wavefront sensor spot position, one row for horizontal and one row for vertical. In Prof. Smythe's example, he has 360 rows. There is one column for each actuator. He has 341 columns.[5]

The number that goes into each place in the matrix is nothing more than how much the Hartmann spot moved when a specific actuator moves. Some people build elaborate computer models[6] and predict what the numbers should be. These people have a lot of time to do it and they trust their answers.

---

[4] It goes by names like *sparse matrix methods*, and others.

[5] 76 trombones and 110 cornets in a band is nothing. How about 40,000 trombones, 120,000 cornets, and 60,000 snare drums?

[6] I suppose supercomputers have supermodels in them?

Other people, like me for instance, build supercomputer models and generate the supermodels.[7] Kenneth still has not met any supermodels. When the computer models are no longer reliable, or the supermodels leave with a rock star, I resort to another method. I forget the computer models. I just turn on the system. I don't start the adaptive optics closed-loop yet. The matrix reconstructor has no numbers in it.

This is how you can do it. Set all the actuator commands to 0, or to where the mirror surface is someplace in the middle of its motion. Note the positions of all the spots. Step through each of the actuators, carefully, one at a time and give each a small command. This is called *poking* the actuator. By looking at all the spots, we can see how they responded. This is called *peeking* at the response.[8] Some of the spots will not move. We just put a 0 into the matrix for that combination of actuator and spot. Some of the spots did move. We record how far they moved and put that number into the matrix. Isn't this fun?

After we have poked all the actuators and peeked at the responses, we have a completely built matrix reconstructor ready to go.[9] The peek-and-poke method is very useful because it calibrates the complicated system each time you turn the system on. Forget the supermodels.

Summary of the tenth chapter

Prof. Smythe has all the pieces necessary to build the optics system. As a final step, he has to program the computer and calculate the reconstructor. Then, when he plugs it in and turns it on, the

---

[7] Age rating: 18 and over. May cause severe eye damage.

[8] The peek-and-poke method is a conventional way of putting something into a computer and then looking to see what actually went in.

[9] Not exactly. There is a small step called matrix inversion. It's like turning the matrix inside-out, upside-down, and backwards, which can be nauseating.

wavefront sensor can see what the atmosphere is doing and how the deformable mirror is correcting for it. He now has an *adaptive* optics system.

Barbara and Kenneth adapt to each other and have a successful marriage.

Chapter 11

Fun with Shining Lasers into Your Eyes

Every laser laboratory in the universe has this sign on the wall:

> Do NOT look into the laser beam
> with your one remaining eyeball.

This is meant to be funny. It is also meant to be a practical warning. Lasers can cause blindness when not used as directed. On the other hand, sometimes it is necessary and even prudent to shine a laser into your eye.

One such time is when you want to take a picture of the back of your eyeball, the retina. On the retina, there are cells that detect light, called photoreceptors. These cells come in two shapes: rods and cones. A lot of eye diseases are related to the rods and cones, so eye doctors would like to get clear images of them.[1] However, the inside of the eyeball is not clear. Oh, maybe for short while, but the eyeball is filled with a fluid called the vitreous humour.[2] It's a clear liquid but it has some variation of index of refraction throughout it. It's like our blobs of atmosphere, but blobs of funny vitreous stuff, instead. And unlike the atmosphere that is floating

---

[1] Eye doctors are called ophthalmologists, doctors of the Ophth.
[2] A funny vitriolic comedian.

around us, this is the turbulence *inside* our eye. In the past, images of the retina were often too blurry to see any detail. To understand the structure and function of the eye, researchers would dissect eyes from deceased folks.

In the 17th century, Isaac Newton used an even more interesting method to find out how the eye responds to outside pressure. He pushed a large needle behind *his own eyeball* to see what happens.[3]

About 15 years ago, a few clever visual science researchers[4] decided to use adaptive optics to correct for the turbulence inside the eye, just like astronomers were doing. One small problem was the source for the wavefront sensor. The retina doesn't naturally glow, so an artificial beacon has to be made from a laser. "You're not going to shine any laser in my eye! What? Are you crazy?" is the patient's normal response. After assuring the patient that the laser power is low enough that it won't cause any damage, the Release of Liability form is signed and things moved along.[5]

Complete adaptive optics systems with wavefront sensors and deformable mirrors have now been used to get strikingly clear images of the rods, cones, the surrounding blood vessels, and other really yucky stuff inside the eyeball. This is a great way to learn about the function and structure of the eye. Way better than sticking needles behind your own eyeballs.

Summary of chapter eleven

Scientists are using adaptive optics to gaze into eyes and get clear images of the retina.

Barbara and Kenneth are just gazing into each other's eyes. Awwwww.

---

[3] Do NOT try this at home. Newton was a genius and you are probably not!
[4] Real citation: J. Liang, B. Grimm, S. Goelz, and J. F. Bille, "Objective measurement of wave aberrations of the human eye with use of a Hartmann-Shack wave-front sensor," *J. Opt. Soc. Am. A* **11**, 1949–1957 (1994).
[5] The "damage threshold" was first determined using the eyes of two rabbits, namely Peter and Volkswagen.

Chapter 12

A Happy Ending

At this point, Leslie B. Smythe turns on his adaptive optics system. He "closes the loop," allowing the deformable mirror to constantly correct for the turbulent atmosphere. He looks at wonderment through his telescope ...

Artist's depiction of wonderment!

... just as a large bird flies into the telescope. The high-tech optical imaging system is now temporarily pointing into the home of Barbara and Kenneth, where they are doing what appears to be ...

Oops.

... having an intellectual conversation about adaptive optics, based solely on the contents of this book.

Bibliography
(Some other books about adaptive optics that aren't nearly as funny as this one)

Introduction to Adaptive Optics, R. K. Tyson, SPIE Press, Bellingham, WA (2000).

Principles of Adaptive Optics, R. K. Tyson, Academic Press, Boston, (1991, 2nd ed. 1997).

Field Guide to Adaptive Optics, R. K. Tyson and B. W. Frazier, SPIE Press, Bellingham, WA (2004).

Adaptive Optics for Astronomical Telescopes, J. W. Hardy, Oxford Univ. Press (1998).

Adaptive Optics, E. Kibblewhite and W. Wild, Wiley, New York (2007).

Adaptive Optics Engineering Handbook, R. K. Tyson, Ed., Marcel Dekker, New York (2000).

Atmospheric Adaptive Optics, V. P. Lukin, *Atmosfernaya Adaptivnaya Optika,* Novosibirsk: Nauka (1986); Translated into English, SPIE Press, Bellingham, WA (1995).

Imaging through Turbulence, M. C. Roggemann and B. Welsh, CRC Press (1996).

Field Guide to Atmospheric Optics, L. C. Andrews, SPIE Press, Bellingham, WA (2004).

Laser Beam Propagation through Random Media, L. C. Andrews and R. L. Phillips, SPIE Press, Bellingham, WA (1998, 2nd ed. 2005).

Introduction to Wavefront Sensors, J. M. Geary, SPIE Press, Bellingham, WA (1995).

Index

Robert K. Tyson is an Associate Professor of Physics and Optical Science at The University of North Carolina at Charlotte. He has a B.S. degree in physics from Penn State University and M.S. and Ph.D. degrees in physics from West Virginia University. Prior to his current academic position, he worked (or occupied an office) in the aerospace industry, first for United Technologies Corporation (West Palm Beach, FL) and then at Schafer Corporation (Chelmsford, MA). Bob Tyson, who enjoys teaching (especially undergraduates), lives in Charlotte, North Carolina with his wife, Peggy, and their cat, Speedway. His hobbies include travel, music, target shooting, and writing politically incorrect satire that is designed to annoy one or more groups of people or, at the least, inspire dynamic, philosophical yelling and screaming.